Stefanie Knauert

Toxicity of pesticides and their mixture to primary producers

Stefanie Knauert

Toxicity of pesticides and their mixture to primary producers

Laboratory and semi-field investigations on the single and joint toxicity of the photosynthesis inhibitors copper, atrazine, isoproturon, and diuron

Südwestdeutscher Verlag für Hochschulschriften

Impressum/Imprint (nur für Deutschland/ only for Germany)
Bibliografische Information der Deutschen Nationalbibliothek: Die Deutsche Nationalbibliothek verzeichnet diese Publikation in der Deutschen Nationalbibliografie; detaillierte bibliografische Daten sind im Internet über http://dnb.d-nb.de abrufbar.
Alle in diesem Buch genannten Marken und Produktnamen unterliegen warenzeichen-, markenoder patentrechtlichem Schutz bzw. sind Warenzeichen oder eingetragene Warenzeichen der jeweiligen Inhaber. Die Wiedergabe von Marken, Produktnamen, Gebrauchsnamen, Handelsnamen, Warenbezeichnungen u.s.w. in diesem Werk berechtigt auch ohne besondere Kennzeichnung nicht zu der Annahme, dass solche Namen im Sinne der Warenzeichen- und Markenschutzgesetzgebung als frei zu betrachten wären und daher von jedermann benutzt werden dürften.

Verlag: Südwestdeutscher Verlag für Hochschulschriften Aktiengesellschaft & Co. KG
Dudweiler Landstr. 99, 66123 Saarbrücken, Deutschland
Telefon +49 681 37 20 271-1, Telefax +49 681 37 20 271-0, Email: info@svh-verlag.de
Zugl.: Basel, Universität Basel, Dissertation, 2008

Herstellung in Deutschland:
Schaltungsdienst Lange o.H.G., Zehrensdorfer Str. 11, D-12277 Berlin
Books on Demand GmbH, Gutenbergring 53, D-22848 Norderstedt
Reha GmbH, Dudweiler Landstr. 99, D- 66123 Saarbrücken
ISBN: 978-3-8381-0037-1

Imprint (only for USA, GB)
Bibliographic information published by the Deutsche Nationalbibliothek: The Deutsche Nationalbibliothek lists this publication in the Deutsche Nationalbibliografie; detailed bibliographic data are available in the Internet at http://dnb.d-nb.de.
Any brand names and product names mentioned in this book are subject to trademark, brand or patent protection and are trademarks or registered trademarks of their respective holders. The use of brand names, product names, common names, trade names, product descriptions etc. even without
a particular marking in this works is in no way to be construed to mean that such names may be regarded as unrestricted in respect of trademark and brand protection legislation and could thus be used by anyone.

Publisher:
Südwestdeutscher Verlag für Hochschulschriften Aktiengesellschaft & Co. KG
Dudweiler Landstr. 99, 66123 Saarbrücken, Germany
Phone +49 681 37 20 271-1, Fax +49 681 37 20 271-0, Email: info@svh-verlag.de

Copyright © 2008 Südwestdeutscher Verlag für Hochschulschriften Aktiengesellschaft & Co. KG and licensors
All rights reserved. Saarbrücken 2008

Produced in USA and UK by:
Lightning Source Inc., 1246 Heil Quaker Blvd., La Vergne, TN 37086, USA
Lightning Source UK Ltd., Chapter House, Pitfield, Kiln Farm, Milton Keynes, MK11 3LW, GB
BookSurge, 7290 B. Investment Drive, North Charleston, SC 29418, USA
ISBN: 978-3-8381-0037-1

„Strömt von der hohen,
steilen Felswand
der reine Strahl,
dann stäubt er lieblich
in Wolkenwellen
zum glatten Fels,
und leicht empfangen
wallt er verschleiernd,
leisrauschend
zur Tiefe nieder."

Johann Wolfgang Goethe

(Gesang der Geister über den Wassern)

Für meine Eltern

Danksagung

Mein Dankeschön geht an alle, die mich während der Dissertationszeit begleitet und in vielerlei Hinsicht unterstützt haben, um diese Arbeit zu einem erfolgreichen Abschluss zu führen.

Mein erster Dank gilt Katja Knauer. Sie hat es mir ermöglicht, dieses interessante Projekt durchzuführen. Ihr ist es mit großem Optimismus immer wieder aufs Neue gelungen, meine Begeisterung zu entfachen und sie hat mich mit ihrer Erfahrung und ihrem Wissen immer voll unterstützt.

Mein Dank geht auch an Herrn Prof. Thomas Boller. Er hat es mir ermöglicht, meine Dissertation am Botanischen Institut fertig zu stellen und die Fakultätsverantwortung übernommen. Mein Dank gilt in diesem Sinne auch Frau Prof. Laura Sigg von der eawag, die sich sehr gern bereit erklärt hat, das Zweitgutachten der Arbeit zu übernehmen.

Ich danke Gabi Thoma vom MGU. Von ihr konnte ich Wertvolles über die Mesokosmen und die Feldarbeit lernen. In diesem Sinne auch vielen Dank an all die anderen, fleißigen Helfer vom MGU, die uns bei den Arbeiten rund um die Mesokosmenstudie unterstützt und somit sehr zum Gelingen des Projekts beigetragen haben. Insbesondere danke ich an dieser Stelle Andrea Leimgruber, Lukas Zimmermann, Helge Abicht, Patrick Schwartz, Heidi Schiffer, Sophia Bloch, Maria á Marca und Oliver Körner.

Mein Dank gilt insbesondere Beate Escher, Juliane Hollender, Heinz Singer und Alfred Lück von der Eawag. Bei Beate bedanke ich mich für die anregenden und fruchtbaren wissenschaftlichen Diskussionen. Juliane, Heinz und Alfred bin ich für die Durchführung der HPLC Analytik der Wasserproben dankbar.

Weiterhin richtet sich mein Dankeschön an Marianne Caroni vom Geographischen Institut und Heinz Hürlimann vom Institut für Umweltgeowissenschaften. Sie haben die Ionenchromatographie und ICP-MS-Messungen der Wasserproben durchgeführt.

Vielen Dank auch an Beat Reber und Tobias Straumann von der Syngenta in Stein/Säckingen. Beide haben uns mit vielerlei nützlichen Gerätschaften versorgt und uns so manches Mal aus der Klemme geholfen.

Mein Dank gilt Giacomo Busco, Kurt Ineichen und Vaclav Mandak vom Botanischen Institut. Sie haben mich rasch und unproblematisch in die Benutzung der Laborgeräte am Botanischen Institut eingewiesen und mich somit sehr bei der Aufarbeitung der Pflanzenproben unterstützt.

Ich möchte mich bei Ursula Dawo von der TU München bedanken, die für uns das Bestimmen und Auszählen der Phytoplanktonproben übernommen hat. Desweiteren danke ich Udo Hommen vom Fraunhofer Institut in Schmallenberg für die Unterstützung bei der statistischen Auswertung der Phytoplankton-Datensätze.

Für ihre hilfreiche Unterstützung und weiterführenden Ratschläge rund um EDV-Fragen möchte ich Rainer Kamber, Lukas Zimmermann, Jürg Oetiker und Roland Preston danken.

Mein herzlicher Dank gilt Felicitas Mäder, Nora Zuberbühler und Erika Roth. Sie haben mir bei administrativen Belangen immer unter die Arme gegriffen und mir damit sehr viel Zeit erspart.

Ein herzliches Dankeschön richtet sich an die gesamte Gruppe des Programm MGU und des Programms Nachhaltigkeit. Die Zeit mit Euch zusammen wird mir immer in guter Erinnerung bleiben.

Im Besonderen möchte ich Florian danken, für seine Geduld, Gelassenheit, Lebenserfahrung und Liebe. Er hat mich immer wieder aufgemuntert und mir die Kraft gegeben, auch die schwierigen Momente der letzten drei Jahre zu meistern.

Diese Arbeit wurde vom Schweizer Bundesamt für Umwelt (Projektnummer 8U01/2006-01/0002 und Projektnummer 2004.H.15a), sowie von der Syngenta Crop Protection AG finanziell unterstützt. Zudem ermöglichte mir ein Stipendium der Freiwilligen Akademischen Gesellschaft Basel, meine Arbeit erfolgreich abzuschließen.

Table of contents

Table of contents i

Zusammenfassung 1

Summary 5

Chapter 1: General Introduction 9
    1.1   Aquatic ecosystems – Importance of macrophytes and phytoplankton 9
    1.2   Pesticides in the aquatic environment 9
        1.2.1   Atrazine, isoproturon, and diuron in the aquatic environment 10
        1.2.2   Copper as pesticide in the aquatic environment 14
    1.3   Monitoring of PSII inhibition in the aquatic ecosystems 15
    1.4   Water quality criteria for pesticides in the aquatic environment 16
    1.5   Aim of this study 21
    1.6   References 22

Chapter 2: The role of reactive oxygen species in copper toxicity to two freshwater
             green algae 29
    2.1   Abstract 29
    2.2   Introduction 29
    2.3   Material and methods 31
        2.3.1   Cu exposure 31
        2.3.2   Test organisms and culture conditions 31
        2.3.3   Detection of ROS formation 32
        2.3.4   Determination of *in vivo* photosynthetic activity 34
        2.3.5   Determination of specific growth rates 35
    2.4   Results 35
        2.4.1   ROS formation 35
        2.4.2   *In vivo* photosynthetic activity 37
        2.4.3   Specific growth rates 39
    2.5   Discussion 39
    2.6   References 42

Chapter 3: Mixture toxicity of three photosystem II inhibitors (atrazine, isoproturon, and diuron) towards photosynthesis of freshwater phytoplankton studied in outdoor mesocosms 49

3.1 Abstract 49
3.2 Introduction 49
3.3 Material and methods 51
    3.3.1 Test chemicals 51
    3.3.2 Determination of test concentrations 51
    3.3.3 Mesocosm experiment 52
    3.3.4 Analytical methods 53
    3.3.5 Effects on photosynthesis 54
    3.3.6 Calculation of toxic units and half-life periods 55
3.4 Results 56
    3.4.1 Herbicide concentrations in the water column 56
    3.4.2 Effects on photosynthesis 58
3.5 Discussion 61
3.6 References 63

Chapter 4: Effects of PSII inhibitors and their mixture on freshwater phytoplankton succession in outdoor mesocosms 69

4.1 Abstract 69
4.2 Introduction 69
4.3 Material and methods 71
    4.3.1 Test chemicals 71
    4.3.2 Mesocosm test site 72
    4.3.3 Application and exposure regime 72
    4.3.4 Chemical and biological sampling 72
    4.3.5 Chemical analysis 73
    4.3.6 Taxonomic determination 73
    4.3.7 Data analysis and statistical analysis 74
4.4 Results 74
    4.4.1 Chemical analysis 74
    4.4.2 Total abundance 75
    4.4.3 Composition of phytoplankton assemblage 75
    4.4.4 Community structure and succession 76

    4.4.5   Number of different taxa and Shannon diversity 81
    4.5   Discussion 83
    4.6   References 85

Chapter 5: Phytotoxicity of atrazine, isoproturon, and diuron to submersed macrophytes in outdoor mesocosms 91
    5.1   Abstract 91
    5.2   Introduction 91
    5.3   Material and methods 93
        5.3.1   Outdoor mesocosm test site 93
        5.3.2   Chemicals and exposure concentrations 93
        5.3.3   Application and sampling 94
        5.3.4   Chemical analysis 94
        5.3.5   Physical/chemical water parameters and nutrients 95
        5.3.6   Macrophytes 95
        5.3.7   Biological endpoints 96
        5.3.8   Statistical analysis 96
    5.4   Results 97
        5.4.1   Physical/chemical water parameters and nutrients 97
        5.4.2   Chemical analysis 97
        5.4.3   Biological endpoints 99
    5.5   Discussion 102
    5.6   References 104

Chapter 6: Concluding remarks and future directions 109
    6.1   Challenges in regulating pesticide mixtures 109
    6.2   Challenges in protecting sustainable freshwater ecosystems 111
    6.3   References 114

Table of contents

Zusammenfassung

Die aquatischen Ökosysteme unserer Erde werden durch die Verunreinigung mit einer Vielzahl von Substanzen anthropogenen Ursprungs bedroht. Pestizide stellen dabei eine wichtige Gruppe von Umweltschadstoffen dar. Sie werden regelmässig in unseren Oberflächengewässern nachgewiesen und treten als Einzelsubstanzen sowie in Mischungen unterschiedlichster Zusammensetzung auf.

Kupfer (Cu) ist aufgrund seiner fungiziden und herbiziden Wirkung ein häufig eingesetztes Pestizid im Obst- und Weinanbau. Die Toxizität von Cu ist generell mit der Konzentration der bioverfügbaren Kupfer-Spezies, d.h. mit der Konzentration der freien Cu^{2+}-Ionen verknüpft. Ein Wirkmechanismus von Kupfer in phototrophen Organismen besteht in der Inhibierung der Photosynthese und steht in Zusammenhang mit der Bildung von reaktiven Sauerstoffspezies (ROS). Bisher ist jedoch noch weitgehend ungeklärt, welche Rolle ROS im Toxizitätsmechanismus von Cu spielen.

Eine Fragestellung dieser Dissertationsarbeit bestand darin, die Rolle von ROS im Toxizitätsmechanismus von Kupfer näher zu untersuchen. Als Testorganismen wurden die Süßwasser-Grünalgenspezies *Pseudokirchneriella subcapitata* und *Chlorella vulgaris* gewählt. ROS-Bildung wurde mittels eines fluorometrischen Assays in den beiden Grünalgen untersucht und mit Kurzzeiteffekten auf die Photosynthese, gemessen als *in vivo* Chlorophyllfluoreszenz, sowie mit Langzeiteffekten auf das Algenwachstum in Zusammenhang gesetzt. Bei einer Exposition gegenüber umweltrelevanten Cu-Konzentrationen von 50 und 250 nM konnte eine vergleichbare Licht- und Zeit-abhängige Zunahme der ROS-Konzentrationen in *P. subcapitata* und *C. vulgaris* bestimmt werden. In *P. subcapitata* führten 250 nM Cu zu einer Reduktion der Photosyntheseaktivität um 12 % während bei *C. vulgaris* keine Effekte auftraten. Diese Ergebnisse weisen darauf hin, dass Unterschiede in den Sensitivitäten der Photosynthese der beiden Grünalgenspezies gegenüber Cu nicht durch Unterschiede in der zellulären ROS-Bildung, sondern eher durch einen Unterschied der Spezies-spezifischen ROS-Abwehrsysteme zu erklären sind. Durch den Einsatz des ROS-Scavengers N-tert-butyl-α-phenylnitron (BPN) konnten die ROS-Konzentrationen in Cu-exponierten Zellen der Spezies *P. subcapitata* auf Kontrollwerte reduziert und die Photosyntheseaktivität vollständig wiederhergestellt werden. Dies deutet darauf hin, dass ROS eine entscheidende Rolle bei der Cu-Toxizität in Grünalgen spielen. In weiteren Experimenten wurde festgestellt, dass ROS in *P. subcapitata* über die Zellmembran in das umgebende Medium abgegeben wurden. Das Verhältnis intra:extra ROS belief sich dabei auf 1:9. ROS-Abgabe könnte somit ein effizienter Mechanismus sein, um die Zelle vor Cu-induzierten oxidativen Schäden zu schützen.

Zusammenfassung

Neben Kupfer sind das Triazin Atrazin und die Phenylharnstoff-Herbizide Isoproturon und Diuron drei häufig auftretende Pestizide in unseren Oberflächengewässern. Im Gegensatz zu Kupfer haben sie einen einzigen spezifischen Wirkmechanismus, der sehr gut untersucht ist. Atrazin, Isoproturon und Diuron inhibieren die Photosynthese, indem sie den photosynthetischen Elektronentransport durch das Photosystem-II (PSII) stören. In Laborstudien konnte gezeigt werden, dass die Mischungstoxizität von Triazinen und Phenylharnstoff-Herbiziden auf Einzel- und Multispeziesebene sehr gut durch das Konzept der Konzentrationsadditivität vorhergesagt werden kann. Bisher fehlen jedoch Studien, die die Anwendbarkeit des Konzepts der Konzentrationsadditivität zur Vorhersage von Mischungseffekten für natürliche Lebensgemeinschaften unter komplexen Umweltbedingungen überprüfen.

Ein weiteres Ziel der vorliegenden Dissertationsarbeit bestand darin, die Anwendbarkeit des Konzepts der Konzentrationsadditivität für natürliche Lebensgemeinschaften zu überprüfen. Hierzu wurden Phytoplanktongemeinschaften und die drei submersen Makrophyten *Myriophyllum spicatum*, *Elodea canadensis* und *Potamogeton lucens* in Süßwasser-Freilandmesokosmen untersucht. Die PSII-Inhibitoren Atrazin, Isoproturon und Diuron dienten als Testsubstanzen und wurden jeweils einzeln und als Mischung appliziert. In den Einzelapplikationen entsprach die Zielkonzentration der HC_{30} („hazardous concentration"), die sich aus einer Speziessensitivitätsverteilung (SSV) ableitet. Die SSV für die drei Substanzen wurden auf Grundlage von Laborwachstumsdaten für verschiedene Algen- und Pflanzenspezies erstellt. In der Herbizidmischung waren alle drei Substanzen zu je 1/3 ihrer HC_{30} vertreten. Unter der Annahme von Konzentrationsadditivität sollte die Mischung dieselben Effekte hervorrufen wie die HC_{30} der Einzelsubstanzen.

Herbizidkonzentrationen und Effekte auf das Phytoplankton und die Makrophyten wurden während eines Zeitraums von 5 Wochen mit konstanter Exposition und einer anschließenden 5-monatigen Nachbehandlungsphase, in der die Herbizide abgebaut wurden, beobachtet. Als Endpunkt wurde die Photosyntheseaktivität des Phytoplanktons und der Makrophyten als direkter Angriffspunkt der Herbizide gewählt. Die Photosyntheseaktivität wurde dabei über *in vivo* Chlorophyllfluoreszenz-Messungen bestimmt. Zudem wurden Effekte auf die Abundanz, Diversität und Spezieszusammensetzung des Phytoplanktons sowie auf das Wachstum der beiden Makrophyten *M. spicatum* und *E. canadensis* untersucht.

Während der ersten fünf Expositionswochen wurden die durchschnittlichen Herbizidkonzentrationen in der Wasserphase erfolgreich konstant zwischen 80 und 120 % der angestrebten Zielkonzentrationen gehalten. Unter Annahme einer Abbaukinetik erster Ordnung entsprachen die Halbwertszeiten der Herbizide in der Nachbehandlungsphase 107 d (Atrazin), 35 d (Isoproturon) und 43 d (Diuron). Am Ende der Studie lagen die Atrazinkonzentrationen noch bei etwa 40 % der Zielkonzentration. Isoproturon und Diuron hingegen konnten nicht mehr signifikant nachgewiesen werden.

Während des Expositionszeitraums der konstanten Konzentrationen erwiesen sich die gewählten Einzelherbizidkonzentrationen als equipotent toxisch bezüglich der Photosyntheseaktivität, der Abundanz und Diversität der Phytoplanktongemeinschaft. Überdies konnte nachgewiesen werden, dass sich die Herbizide konzentrationsadditiv verhielten, da die Effekte der Mischung auf Photosyntheseaktivität, Abundanz und Diversität des Phytoplanktons denen der Einzelsubstanzen vergleichbar waren. Als Folge unterschiedlicher Sensitivitäten weniger Algenspezies gegenüber den Herbiziden wurde eine zum Teil anders geartete Sukzession der verschieden behandelten Gemeinschaften im Verlauf der Studie beobachtet. Die mit Diuron und Isoproturon behandelten Algengemeinschaften unterschieden sich kurze Zeit nach Ende der konstanten Expositionsphase in Bezug auf Photosyntheseaktivität, Abundanz, Diversität und Spezieszusammensetzung nicht mehr wesentlich von den unbehandelten Phytoplanktongemeinschaften. Dies konnte zum einen auf den raschen Rückgang der Herbizidkonzentrationen im Wasser zurückgeführt werden. Zum anderen ließ sich jedoch eine Toleranzentwicklung bei einzelnen Algenspezies als Ursache für die beobachtete Wiedererholung nicht ausschließen. Die Phytoplanktongemeinschaften, welche mit Atrazin bzw. mit der Herbizidmischung behandelt wurden, waren bis zum Ende des Nachbehandlungszeitraumes in ihrer Photosyntheseaktivität beeinträchtigt und unterschieden sich auch wesentlich in der Spezieszusammensetzung von den unbehandelten Algengemeinschaften. Die anhaltende Exposition gegenüber Atrazin, das eher langsam abgebaut wurde, führte hier zu einer anders gerichteten Sukzession der Phytoplanktongemeinschaft.

In den drei submersen Makrophyten wurde eine Inhibierung der Photosyntheseaktivität nur in einem kurzen Zeitfenster an Tag 2 und 5 direkt nach der ersten Applikation gemessen. Dies deutet darauf hin, dass sich die Makrophyten sehr rasch an den chemischen Stress adaptierten. Die beobachteten Kurzzeiteffekte der Herbizide und ihrer Mischung auf die Photosyntheseaktivität erklärten zudem, dass keine signifikanten Effekte auf das Wachstum von *M. spicatum* und *E. canadensis* gefunden werden konnten. Darüber hinaus erwiesen sich die Einzelherbizidkonzentrationen als equipotent toxisch in *M. spicatum*, weil sie vergleichbare Effekte auf die Photosyntheseaktivität hervorriefen. Da auch die Mischung zu einer den Einzelsubstanzen vergleichbaren Inhibierung der Photosyntheseaktivität in *M. spicatum* führte, konnte Konzentrationsadditivität der drei Herbizide für diese Makrophyte gezeigt werden. In *E. canadensis* und *P. lucens* induzierten die HC_{30} von Atrazin, Isoproturon und Diuron keine vergleichbaren

Zusammenfassung

Effekte auf die Photosynthese und waren somit nicht equipotent toxisch. Aus diesem Grund konnte hier keine weiterführende Aussage in Bezug auf die Konzentrationsadditivität der Herbizide in diesen Makrophyten getroffen werden.

Mit dieser Mesokosmen-Studie konnte am Fallbeispiel von drei PSII-Inhibitoren die Anwendbarkeit des Konzepts der Konzentrationsadditivität für eine natürliche Algengemeinschaft und für die Makrophyte *M. spicatum* unter Freilandbedingungen bestätigt werden. Die hier gewonnenen Erkenntnisse können einen Beitrag zur aktuellen Diskussion über die Berücksichtigung von Mischungen bei der Definition der Gewässerqualität leisten, um die nachteiligen Auswirkungen von Pestizidgemischen auf die aquatischen Lebensgemeinschaften zu begrenzen und damit einen nachhaltigen Schutz der aquatischen Ökosysteme zu gewährleisten.

Summary

The earth's aquatic ecosystems are threatened by the contamination with a multitude of anthropogenic chemical pollutants. Pesticides are one important group of environmental contaminants. They are frequently detected in our surface waters and occur as single substances and in mixtures of various compositions.

Copper (Cu) is often used as fungicide and herbicide in orcharding and viniculture. Cu toxicity is generally linked to the bioavailable fraction, i.e. to the concentration of the free Cu^{2+}. In phototrophic organisms, one toxic mode of action of Cu is due to the inhibition of photosynthesis. Phytotoxicity of Cu was also found to be related to the generation of reactive oxygen species (ROS). However, until now, it is not clear, whether ROS are a mere consequence of Cu toxicity or the primary cause.

One objective of this dissertation thesis was thus to investigate the role of ROS in the toxicity of Cu to phototrophic organisms to gain a better understanding of its toxicity mechanism. The two freshwater green algal species *Pseudokirchneriella subcapitata* and *Chlorella vulgaris* were chosen as test organisms. Cu-induced ROS formation was investigated in relation to short-term effects on photosynthetic activity and long-term effects on growth of *P. subcapitata* and *C. vulgaris*. Photosynthetic activity was determined as *in vivo* chlorophyll fluorescence. Exposure to 30 nM and 300 nM Cu resulted in a light and time dependent increase in ROS concentrations in *P. subcapitata* and *C. vulgaris*. The potential of Cu to induce ROS was comparable in both algae but the effect on photosynthesis differed with 300 nM Cu leading to a 12 % reduction of photosynthetic activity in *P. subcapitata* but not *C. vulgaris* after 24 h. This indicates that species-specific sensitivities were not caused by differences in ROS content but more likely resulted from differences in each ROS defence systems. The ROS scavenger N-tert-butyl-α-phenylnitrone (BPN) diminished Cu induced ROS production to control levels and completely restored Cu inhibiting effects on photosynthetic activity of *P. subcapitata*. This implies that ROS may play a primary role in the mechanism of copper toxicity to photosynthesis in algal cells. Further experiments revealed a time-dependent ROS release process across the plasma membrane. More than 90 % of total ROS were determined to be extracellular in *P. subcapitata*, indicating an efficient way of cellular protection against oxidative stress.

Besides Cu, the triazine atrazine and the phenyl urea herbicides isoproturon and diuron are frequently detected in our surface waters. In contrast to Cu, phytotoxicity of these three pesticides is due to one specific mode of action which has been thoroughly investigated. Atrazine, isoproturon, and diuron inhibit photosynthesis by interrupting electron transport through photosystem II (PSII).

Summary

Laboratory studies demonstrated that mixture toxicity of triazine and phenylurea herbicides to single species and communities is predictable by the concept of concentration addition. However, there is a lack of studies that verify the applicability of these concepts for natural communities exposed under complex environmental conditions.

A further objective of this thesis was thus to verify if the concept of concentration addition can be applied also for natural communities exposed under realistic environmental conditions. Therefore, a phytoplankton community and the three submersed macrophytes *Myriophyllum spicatum, Elodea canadensis,* and *Potamogeton lucens* were studied in freshwater outdoor mesocosms. The three PSII inhibitors atrazine, isoproturon, and diuron were chosen as test substances and applied as single substances and in a mixture. In the single treatments the 30 % hazardous concentrations (HC_{30}) of the three substances derived from species sensitivity distribution (SSD) curves were used. The SSD curves were established on the basis of EC_{50} growth inhibition data obtained from laboratory tests with different algal and plant species. The herbicide mixture comprised one third of the HC_{30} of each individual herbicide. If the concept of concentration holds true the herbicide mixture was expected to elicit the same toxic effects as the HC_{30} of three herbicides alone.

Herbicide concentrations and effects on phytoplankton and macrophytes were investigated during a five-week period of constant concentrations and a subsequent five-month post-treatment period when the herbicides dissipated from the water phase. Photosynthetic efficiency of phytoplankton and the three macrophytes was selected as an endpoint directly linked to the mode of action of the three test substances. Moreover, effects on abundance, diversity, and species composition of phytoplankton as well as on growth of the two macrophytes *E. canadensis* and *M. spicatum* were examined.

In the period of constant concentrations averaged herbicide water concentrations were determined to be in the range of target concentrations ± 20 %. In the post-treatment period the dissipation of the herbicides was described by first order kinetics. Half-lives corresponded to 107 d for atrazine, 35 d for isoproturon, and 43 d for diuron. At the end of the experiment atrazine concentrations in the water phase still reached approximately 40 % of the target concentration whereas isoproturon and diuron had nearly completely disappeared.

In the constant exposure period the single herbicides were shown to be equitoxic due to comparable effects on photosynthetic efficiency, abundance, and diversity of phytoplankton. Furthermore, the herbicides were found to act concentration additive since the effects of the mixture on photosynthetic efficiency, abundance, and diversity were similar to those of the single substances. Because of different sensitivities of a few algal species towards the herbicides, species composition of the communities in the various treatments developed differently during the post-treatment period. Diuron and isoproturon treated algal communities did not differ considerably from the untreated communities concerning photosynthetic efficiency, diversity, and species composition already a short time after the end of the period of constant exposure. This might be linked to the rapid decrease in herbicide concentrations in the water phase. However, tolerance of single algal species towards the herbicides might have also contributed to the recovery of diuron and isoproturon treated phytoplankton. Photosynthetic efficiency and species composition of atrazine and mixture treated phytoplankton was found to be adversely affected and considerably different compared to the untreated communities until the end of the post-treatment period. Continuous exposure to persisting atrazine concentrations resulted in a different succession of phytoplankton in these two treatments.

In the three submersed macrophytes inhibition of photosynthesis was determined in a short time window from day 2 to 5 after first application only indicating a rapid adaptation of the macrophytes towards herbicide stress. The observed short term effects of the herbicides and their mixture on photosynthesis might also explain that growth of *M. spicatum* and *E. canadensis* was not affected. In addition, the single herbicide concentrations turned out to be equitoxic in *M. spicatum* since they elicited similar effects on photosynthetic efficiency of this macrophyte. Concentration addition of atrazine, isoproturon, and diuron could also be verified for this macrophyte since the mixture inhibited photosynthetic efficiency comparable to the single substances. In *E. canadensis* and *P. lucens* the HC_{30} of atrazine, isoproturon, and diuron did not stimulate a similar inhibition of photosynthesis and were thus found to be not equitoxic. For this reason, any conclusions on concentration addition of the herbicides in these two macrophytes could not be drawn.

This case study confirmed the applicability of the concept of concentration addition for three PSII inhibitors when considering their effects on a natural algal community and on the macrophyte *M. spicatum* under environmental conditions. The results can thus contribute to the current discussion concerning the incorporation of mixture toxicity in the regulation of surface water quality to adequately protect aquatic communities from pesticide impact and to guarantee a sustained management of the aquatic ecosystems.

Summary

Chapter 1

General introduction

1.1 Aquatic ecosystems – Importance of macrophytes and phytoplankton

Freshwater ecosystems such as lakes, streams, rivers, or wetlands support an enormous biodiversity of plant and animal life. Macrophytes play an important role in freshwater ecosystems by providing nutrient cycling, primary productivity, food, and habitat for other organisms. They are acting as ecological engineers, having great impact on the physical properties of the aquatic habitat. Especially in oligotrophic ponds and lakes as well as in streams and wetland communities, submersed, floating, and emergent macrophytes are essential to harbor diverse animal communities. As the aquatic plants, phytoplankton is contributing to primary production and nutrient cycling. Phytoplankton forms the basis of the pelagic food web (Wetzel 2001, Reynolds 2006). For these reasons, primary producers such as macrophytes and phytoplankton are especially addressed in the sustainable management and protection of freshwater ecosystems (EU Water Framework Directive, 2000/60/EC).

1.2 Pesticides in the aquatic environment

Due to anthropogenic activities, the freshwater ecosystems throughout the world have been increasingly contaminated with a multitude of industrial and natural chemical compounds (Schwarzenbach et al. 2006). In particular, increasing environmental awareness has generated concerns regarding the impact of pesticides on aquatic ecosystems. Pesticides have become and will continue to be an integral part of modern crop protection in an intensive agriculture satisfying consumption needs and food supply for the increasing world's human population (Streibig and Kudsk 1993). Compared to other xenobiotics, pesticides are unique chemical stressors since they are designed to have biological activity and are intentionally placed into the environment in large amounts. Pesticides are mainly grouped based on their target pests, e.g. herbicides, insecticides, nematicides, rodenticides, acaricides or fungicides. In a more detailed framework, pesticides are grouped into classes of compounds that have similar chemical structures and modes of toxic action. Among herbicides, for instance, we find inhibitors of photosynthetic electron transport (e.g. bipyridillum, triazines, triazinones, phenylureas, uracils), inhibitors of branched chain amino acid biosynthesis (e.g. sulfonylureas, imidazolinones), inhibitors of aromatic amino acid biosynthesis (e.g. glyphosate), inhibitors of mitosis (e.g. dinitroanalines), inhibitors of fatty acid biosynthesis (e.g. cyclohexanediones), auxinic herbicides (e.g. 2, 4 dichlorophenoxyacetic acid), and others. Most herbicides are, in contrast to other pesticides, polar substances characterized by high water solubility and low sorption coefficients allowing for rapid uptake into target plants (Johnson and Ebert 2000). Such physicochemical features render them susceptible to transfer from the deliberately treated areas into the non-target aquatic environment. Estimates indicate that the average herbicide loss is around 1 % of the applied volume (Wauchope 1978, Kreuger 1998, Carter 2000). Traces of herbicides, and mixtures of them, are frequently detected in aquatic ecosystems in agricultural landscapes but also in urban catchments (Kreuger 1998, Müller et al. 2002, Irace-Guigand et al. 2004, Leu et al. 2004). In particular, the triazine atrazine and the two phenylurea herbicides diuron and isoproturon are often occurring in surface waters as single substances or in mixture (Fig. 1.1) (Nitschke and Schüssler 1998, Graymore et al. 2001, Field et al. 2003, Kotrikla et al. 2006).

Fig. 1.1 Structural formula of atrazine, isoproturon, and diuron

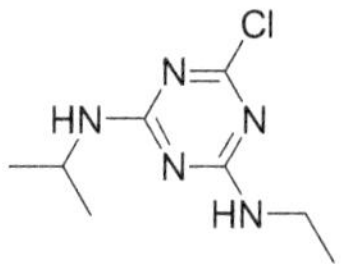

Atrazine

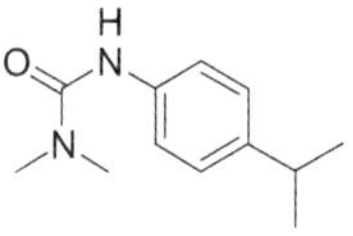

Isoproturon

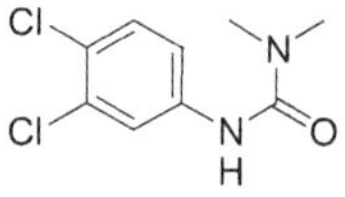

Diuron

1.2.1 Atrazine, isoproturon, and diuron in the aquatic environment

Atrazine was developed in the 1950s by the Geigy Chemical Company of Basel, Switzerland. Since that time, it has been widely used in agricultural applications throughout the world. Physicochemical properties of atrazine (low vapor pressure, low Henry's law constant, moderate water solubility, small K_{OC}/K_{OW}) (Table 1.1) and the marginal degradation via biotic and abiotic pathways results in a relatively high persistence and mobility of this substance in the water phase. Thus, application of atrazine was prohibited in many countries. In Germany, for instance, atrazine was banned in 1991. In others countries, such as Switzerland, the use of atrazine has been restricted to some minor applications.

Isoproturon and diuron have similar physicochemical features compared to atrazine (Table 1.1). However, the two phenylureas are substantially degraded by microorganisms under field conditions whereas non-biological degradation rather insignificantly contributes to their dissipation (Vroumsia et al. 1996, Sørensen et al. 2001, Field et al. 2003).

Table 1.1 Physicochemical properties of atrazine, diuron and isoproturon (according to Mackay et al. 2006, Tomlin 2006)

	Atrazine	Diuron	Isoproturon
IUPAC name	1-chloro-3-ethyl-amino-5-isopropyl-amino-2,4,6-triazine	3-(3,4-dichlorophenyl)-1,1-dimethylurea	3-(4-isopropylphenyl)-1,1-dimethylurea
CAS registry number	1912-24-9	330-54-1	34123-59-6
Molecular weight	215.7 g/mol	233.1 g/mol	206.3 g/mol
Water solubility	33 mg/L (22 °C)	37 mg/L (25 °C)	65 mg/L (22 °C)
Vapor pressure	3.9×10^{-2} mPa (25 °C)	1.1×10^{-3} mPa (25 °C)	3.2×10^{-3} mPa (20 °C)
Henry´s Law Constant	1.5×10^{-4} Pa m^3 mol^{-1}	5.2×10^{-5} Pa m^3 mol^{-1}	1.5×10^{-5} Pa m^3 mol^{-1}
Octanol/Water Partition Coefficient, log K_{OW}	2.5 (25 °C)	2.9 (25 °C)	2.5 (20 °C)
Sorption Partition Coefficient, log K_{OC}	1.95-2.19	2.6	1.72-2.8

Several reviews reported on the ecological and ecotoxicological relevance of atrazine for the aquatic environment (Huber 1993, Solomon et al. 1996, Graymore et al. 2001). Atrazine possesses a relatively low toxicity since concentrations up to 20 µg/L are not assumed to cause any permanent damage to aquatic ecosystems (Huber 1993). Ecotoxicity data of the two phenylurea herbicides isoproturon and diuron are available for a number of algal species but are rather rare for macrophytes. Generally, they have been found to be more toxic to a number of phototrophic organisms than the triazine (compilation of ecotoxicity data in Chèvre et al. 2006, supporting information).

In phototrophic organisms, atrazine, isoproturon and diuron share the same toxic mode of action. They inhibit photosynthesis acting as photosystem II inhibitors (Trebst 1987). Photosynthesis is one of the most important biochemical pathways since nearly all life on earth depend on it. Approximately 45 % of the photosynthesis on Earth occurs in aquatic ecosystems (Falkowski and Raven 1994). Photosynthesis is an oxidation-reduction process where light energy is converted to chemical bond energy that is stored in the form of organic carbon compounds. Plants, algae, and prokaryotic cyanobacteria utilize a water-cleaving photosynthetic reaction:

$$6\ CO_2 + 12\ H_2O \xrightarrow{h\nu} C_6H_{12}O_6 + 6\ O_2 + 6\ H_2O$$

where water is oxidized and the released electrons are energized and ultimately transferred to electron acceptor carbon dioxide, yielding carbohydrates and oxygen.

The photosynthetic process involves two phases. The photosynthetic light reactions, where O_2, ATP and NADPH are produced, are organized on thylakoid membranes of specialized organelles, the chloroplasts. The carbon-fixing reactions, which consume the ATP and NADPH of the light reaction to produce carbohydrates by reduction of CO_2, are generally localized in the aqueous phase of the chloroplast stroma. Light-driven photosynthetic electron transport involves three membrane-spanning protein complexes, namely the photosystem II (PSII) complex, the cytochrome b_6/f complex and the photosystem I (PSI) complex, as well as a number of other mobile and integral thylakoid membrane or soluble components (e.g. plastocyanin, ferredoxin) that cooperate in the light-driven transfer of electrons from water to $NADP^+$ resulting in the production of O_2 and NADPH (Buchanan et al. 2000).

The PSII complex is composed of the manganese containing oxygen evolving (water-splitting) complex, a reaction center complex, and the light-harvesting chlorophyll antenna proteins. It is a supramolecular entity made up of at least 23 different polypeptides including catalytic, regulatory, and structural subunits as well as several chlorophyll-binding proteins (Andersson and Styring 1991). The reaction center of PSII is composed of a heterodimer of two integral membrane proteins, named D1 and D2 which bind electron transfer prosthetic groups such as P680, pheophytin, and plastoquinon that are involved in the primary photochemistry and the water oxidation processes (Michel and Deisenhofer 1986, Nanba and Satoh 1987). PSII functions as a light-dependent water-plastoquinon oxidoreductase. The light harvesting pigments associated with PSII transfer excitation energy to P680, the chlorophyll *a* dimer in the PSII reaction center. Subsequently, charge separation takes place and an excited electron is released from P680 and transferred to pheophytin and from there to the primary quinon-type acceptor Q_A. The charge separation creates the very highly oxidant P^+680 which is neutralized by receiving an electron from the secondary donor Z, a tyrosine residue of the D1 protein. An electron derived from the splitting of water reduces Z^+. The electron is then transmitted from Q_A to the quinon Q_B. Q_A and Q_B occupy special binding niches at the D1 and D2 protein of the PSII reaction center. In contrast to Q_A, which is rather tightly bound to the PSII reaction center complex, Q_B can dissociate from the PSII complex and function as a mobile electron carrier in the lipid bilayer. After receiving a second electron, Q_B binds two protons from the lumen side of the thylakoid membrane and merges into the plastoquinon/plastohydroquinon (PQ) pool (Hansson and Wydrzynski 1990).

PSII electron transport inhibitors bind to the Q_B-binding niche on the D1 protein. There they act as non-reducible analogs of plastoquinon (Fig. 1.2). The Q_B binding to D1 involves hydrogen bonding between the carbonyl oxygens of plastoquinon and the amide backbone of His 215 and the hydroxyl of Ser 264. PSII herbicides interact with D1 due to hydrogen bonds, van der Waals forces, and hydrophobic interactions. According to a concept of overlapping, but not identical, binding sites on the D1 protein, two different groups of PSII herbicides have been distinguished: the "classical" urea/triazine type inhibitors (e.g. triazines, phenylureas, triazinones, biscarbamates) that strongly interact with Ser 264 and the phenol type inhibitors which interact with His 215 (e.g. nitro-phenols, azaphenanthrenes, hydroxypyridines). The reason why such a diversity of chemical families binds to the D1 protein may be due to the dual binding roles of the D1 as it has to interact with non-reduced as well as singly reduced plastoquinon (Fuerst and Norman 1991, Trebst 1987).

The primary effect of the displacement of plastoquinon from its Q_B binding niche by PSII inhibitors is a block of electron flow through PSII resulting in the inhibition of photosynthetic oxygen evolution, NADP reduction and photophosphorylation. However, the transfer of excitation energy from chlorophyll molecules to PSII reaction centre is indirectly interrupted, too. Excited chlorophyll molecules spontaneously form triplet chlorophyll which can react with molecular oxygen yielding in the formation of singlet oxygen and other reactive oxygen species. Therefore, the phytotoxicity of PSII herbicides is mainly due to photooxidation of the photosynthetic apparatus components embedded in the thylakoid membranes, i.e. due to lipid peroxidation, degradation of chlorophyll and oxidative protein damage. In addition, binding of PSII inhibitors interferes with the degradation of D1 by a protease occurring during normal functioning of the PSII reaction center. As a consequence, the damaged D1 protein cannot be replaced (Merlin 1997).

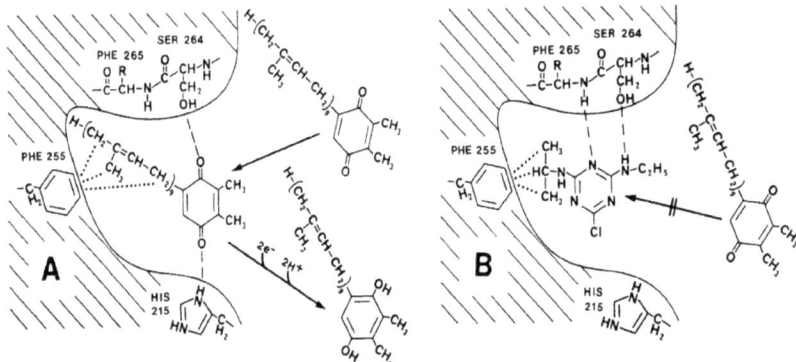

Fig. 1.2 Schematic figure of the plastoquinon/herbicide binding pocket of the D1 protein (adapted from Fuerst and Norman (1991)).

1.2.2 Copper as pesticide in the aquatic environment

In contrast to organic pesticides, heavy metals occur naturally in the environment, and several of them are essential components of ecosystems. Metals such as copper and zinc are essential to life, whereas others such as lead and mercury are not known to perform any useful biochemical function (Allan 1997). Increased metal concentrations in the aquatic ecosystems can occur naturally but can also be related to different anthropogenic sources, e.g., agricultural fungicide and herbicide runoff (Macfarlane and Burchett 2001).

The bioavailability of metal ions to marine and freshwater algae, as well as their toxic effects, strongly depends on the chemical speciation (Sunda 1988, Knauer et al. 1997). Biological effects have been shown to be related to the free metal ion concentration. The mechanism underlying toxicity of metals is not always related to one specific mode of action. Copper, for instance, interferes with different metabolic pathways including photosynthesis, chlorophyll synthesis, fatty acid metabolism or carbohydrate synthesis. The most important effect of copper on plants and algae is associated with the inhibition of photosynthesis. Inhibition sites range from photosynthetic electron transport and photophosphorylation to dark reactions (Fernandes and Henriques 1991).

The mechanism of copper toxicity to photosynthetic electron transport has been widely investigated whereby PSII was found to be more sensitive to copper inhibition than PS I (Cedeno-Maldonado et al. 1972). However, the precise location of the copper inhibitory binding site is still not exactly known. On the one hand, scientists have argued that copper-induced PSII inhibition is located at the acceptor side of PSII whereas other studies suggested that copper impairs PSII electron transport on the donor side (Yruela et al. 1996, Pätsikkä et al. 2001).

Copper-induced inhibition of photosynthesis was found to be strongly related to the production of reactive oxygen species since a number of studies reported on the activation of the antioxidant defense system as well as on an increase in the levels of lipid peroxidation and protein carbonylation (Mallick 2004, Devi and Prasad 2005). However, neither study tried to link intracellular levels of reactive oxygen species to effects on photosynthesis. Thus, one aim of this thesis was to elucidate the role of reactive oxygen species in the mechanism of copper phytotoxicity.

1.3 Monitoring of PSII inhibition in the aquatic ecosystems

The determination of *in vivo* chlorophyll fluorescence has been applied in a number of studies to assess the toxicity of PSII inhibitors to freshwater single species and communities (El Jay et al. 1997, Snel et al. 1998, Dorigo and Leboulanger 2001, Fai et al. 2007) and has been suggested as useful tool to monitor herbicides in surface water and groundwater samples (Conrad et al. 1993, Koblizek et al. 1998, Bengtson Nash et al. 2005). In the last years, this method has been developed as an alternative practical tool for non-intrusive assessment of *in vivo* photosynthesis in intact plant leaves, algae, and isolated chloroplasts (Schreiber et al. 1994).

Chlorophyll exists in the form of pigment protein complexes embedded in the thylakoid membrane of the chloroplasts. Excitation energy is funneled into the reaction centers (P680 and P700) of PSI and PSII. De-excitation occurs not only via photochemical energy conversion but also due to heat dissipation and fluorescence emission (Fig. 1.3). Fluorescence emission originates almost completely from PSII at room temperature whereas PSI is essentially non-fluorescent. The excitation transfer to PSI may be considered as an additional competitive pathway of de-excitation. The rate of fluorescence emission, F, is thus proportional to the absorbed light flux, I_a and to the quantum yield of the fluorescence ΦF and can be described by the following equation (Krause and Weis 1991):

$$F = I_a \times \Phi F = I_a \times k_F/(k_F + k_D + k_T + k_P)$$

where k denotes the rate constants of the following processes:

k_F: fluorescence emission

k_D: thermal deactivation

k_T: transfer of excitation energy to non-fluorescent pigments (e.g. to antennae of PSI)

k_P: photochemical reaction.

During photosynthesis running at high efficiency, the photochemical reaction must be strongly favored over the competing processes ($k_P \gg k_F + k_D + k_T$). Otherwise, valuable energy required for the reduction of CO_2 to carbohydrate would be lost. The interaction of phytotoxic substances, e.g. PSII inhibitors, with the photosynthetic process causes characteristic changes of chlorophyll fluorescence parameters.

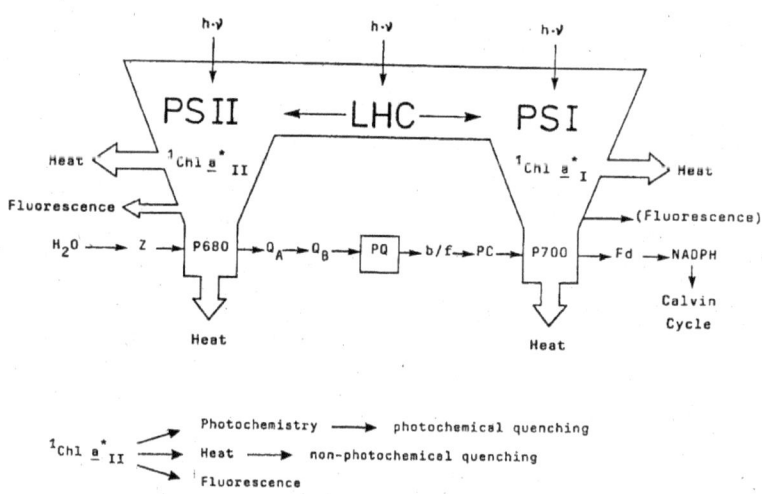

Fig. 1.3 Schematic illustration of primary energy conversion in photosynthesis governing chlorophyll fluorescence yield. LHC – light harvesting complex, Z – particular tyrosine residue of D1 protein, Q_A, Q_B, PQ – plastoquinon molecules, b/f – cytochrome bf6, PC – plastocyanin, Fd – ferredoxin (adapted from Schreiber et al. 1994).

1.4   Water quality criteria for pesticides in the aquatic environment

During the last years public concern has focused on the adverse impact of pesticides on aquatic ecosystems. A number of European and North American countries have thus defined water quality criteria (WQC) for single pesticides in surface waters (Stephan et al. 1985, Zabel and Cole 1999, Crommentuijn et al. 2000, Jahnel et al. 2001).

WQC were first developed in the 1950s and have emerged as one of the primary tools for assessing and managing the biological integrity of surface water. WQC are defined as numerical concentrations or narrative statements and are recommended as levels that should result in negligible risk to biota, their functions, or any interactions that are important to sustain the health of the aquatic ecosystem. WQC are often equivalent to predicted no-effect concentrations which are estimated by finding the lowest reliable effect concentrations and applying a safety factor to consider various uncertainties, such as differences in species sensitivities, acute to chronic ratios or laboratory to field extrapolations. In the Netherlands, WQC are calculated on the basis of hazardous concentrations derived from species sensitivity distribution (SSD) curves (Crommentuijn et al. 2000). An SSD, constructed on the basis of single species data (e.g. no observed effect concentrations or effect concentrations), is a statistical function describing the variation in toxicity of a certain compound among a set of species. The WQC in the Netherlands are based on the HC_5 which represents a concentration that should protect 95 % of the aquatic species. In Switzerland, WQC for pesticides are still arbitrarily defined without any ecotoxicological justification. The current criterion for each individual pesticide accounts for 0.1 µg/L (Chèvre et al. 2006).

Generally, an important shortcoming in the procedure of WQC setting is that mixtures of chemicals as they are occurring in the aquatic environment are currently not considered. Approaches for the assessment of mixture toxicity to determine water quality standards cannot simply rely on experimentally based hazard assessment since testing of all conceivable combinations of substances is not feasible. However, if individual mixture components can be identified, the overall toxicity of the mixture can be predicted from the toxicities of the individual constituents.

There are three basic classes of mixture toxicity: additivity, synergism and antagonism (Calamari and Alabaster 1980, Calabrese 1991). Additive joint action of toxicants occurs when the toxicity of a mixture corresponds to the expected toxicity of the sum of the toxicities of the individual mixture components. Two or more chemicals act synergistic when the toxicity of the mixture is greater than additivity. Finally, antagonistic interaction occurs when the toxicity of the mixture is smaller than assumed if the mixture was additive.

Methods to assess the joint action of mixture components primarily are based on the mathematical groundwork of Bliss (1939). Hewlett and Plackett (1952) expanded the scheme developed by Bliss and postulated four possible types of interactions occurring between chemical components of a mixture (Table 1.2).

Table 1.2 Possible types of interactions that can occur between chemical components of mixtures (according to Hewlett and Plackett 1952).

	Similar Joint Action	Dissimilar Joint Action
Non-Interactive	Simple similar (concentration addition)	Independent (response addition)
Interactive	Complex similar	Dependent

In the non-interactive types of joint action chemicals are assumed to not affect the toxicity of each other. Depending on the toxic mode of action, two different types of non-interactive joint action can be distinguished: simple similar (concentration addition) and independent joint action.

The basic idea of the concept of concentration addition is that all components share an identical mechanism of action in the exposed organism. The concept of concentration addition can be traced back to the early work of Frei (1913) and Loewe and Muischnek (1926). For a multi-component mixture of n substances, it is defined by

$$\sum_{i=1}^{n} \frac{c_i}{EC x_i} = 1$$

(Berenbaum 1985). In this equation, c_i are the concentrations of the individual substances i present in a mixture with a total effect of x %. EC_{xi} are the equivalent effect concentrations of the single substances i, i.e. those concentrations that alone would cause the same quantitative effect x as the mixture. Quotients c_i/EC_{xi} express the concentrations of mixture components as fractions of equi-effective individual concentrations and have been termed *toxic units* (Sprague 1970). If concentration addition as described by the mentioned equation holds true, a mixture component can be replaced totally or in part by an equal fraction of an equi-effective concentration of another without altering the overall effect of the mixture. Or in other words, the total effect of the mixture is expected to remain constant as long as the sum of the toxic units remains constant.

Independent joint action (also referred to as response addition) deals with substance mixtures where the individual components interact with different molecular target sites but contribute to a common effect (e.g., death). For example, if a metal or a herbicide co-occurs with an insecticide, the modes of toxic action may differ. Under independent action, the components in the mixture are assumed to behave independently of one another, so that the organism's response to the first pesticide is the same whether or not the second is present. Independent action indicates that the toxicity of the compounds is predicted to occur based on simple probability statistics. If a concentration of compound A generally kills 25 % of the organisms and a concentration of compound B kills 25 % of the organisms, then the two concentrations of compounds A and B combined would result in their individual effects added together, minus that proportion of the population in which sensitivities overlap. The combination of these two concentrations would kill 43.8 % of the population. The following equation describes independent action for a multi-component mixture (Faust et al. 2003):

$$E(c_{mix}) = 1 - \prod_{i=1}^{n}(1 - E(c_i))$$

where n is the number of compounds in the mixture, $E(c_i)$ is the effect elicited by the ith compound and $E(c_{mix})$ is the predicted joint effect of the mixture.

Complex similar and dependent joint actions are types of mixture toxicity where at least one component in the combination affects the biological activity of at least one other component in the mixture. The biological activity can be altered by influencing the compound's uptake and metabolism or by changing its physiological action. In the case of complex similar joint action the components in the mixture have the same toxic mode of action whereas in the case of dependent joint action the mixture components share different sites of action.

Early studies investigating the toxicities of mixtures to aquatic organisms dealt with the effects of heavy metals and industrial organic pollutants on survival of fish and daphnids (EIFAC 1987). In the last 20 years, combination effects of other types of toxicants, such as pharmaceuticals and pesticides, on various aquatic species have been intensively investigated (Altenburger et al. 2000, Backhaus et al. 2000, Silva et al. 2002, Scholz et al. 2006). Concentration addition was shown to predict mixture effects of similar acting herbicides, such as triazines (Faust et al. 2001, Drost et al. 2003), phenylureas (Backhaus et al. 2004a), chloroacetanilides (Junghans et al. 2003a) or sulfonylureas (Junghans et al. 2003b) on growth of single algal species and duckweed.

Little is known about the effects of pesticide mixtures on multi-species systems (Vighi et al. 2003). Only a small number of studies focused on the application of mixture toxicity approaches on a community level (Backhaus et al. 2004b, Arrhenius et al. 2004, Arrhenius et al. 2006). Arrhenius et al. (2004), for instance, investigated the joint toxicity of 12 phenylurea herbicides on photosynthesis of marine periphyton and epipsammon communities. These algal communities were sampled in the field but exposed under a laboratory test design. From their findings they concluded that the concept of concentration addition applies also at the community level of algal testing when a physiological short-term effect indicator is used that matches the toxic mode of action of the substance.

Table 1.3 Studies addressing mixture toxicity of different types of pesticides on freshwater model ecosystems.

Pesticides in the mixture	Reference
Insecticide + Herbicide:	
Esfenvalerate + Atrazine	Fairchild et al. 1994
Cypermethrin + Methsulfuron methyl	Wendt-Rasch et al. 2003
Insecticide + Insecticide:	
Chlorpyrifos + Lindane	Van den Brink et al. 2002
Fungicide + Insecticide + Herbicide + Herbicide:	
Fluazinam + Lambda-cyhalothrin + Asulam + Metamitron	Wendt-Rasch et al. 2004
Herbicide + Herbicide + Herbicide:	
Atrazine + Diuron + Metolachlor	Hartgers et al. 1998

Mixture toxicity studies as described above were all conducted under laboratory conditions. However, mixture toxicity in the field might be completely different than in a laboratory setting due to the complex environmental and ecological factors, such as chemical partitioning, bioavailability or species-species interactions.

Experimental aquatic ecosystems have become widely used tools in ecotoxicology, because they allow a greater degree of control, replication, and repeatability than can be achieved in natural ecosystems. In the past years, some studies addressed mixture toxicity of different types of pesticides on freshwater model ecosystems simulating realistic environmental exposure scenarios (Table 1.3). Generally, these studies discussed the observed effects in the light of common risk assessment procedures. However, none of these model ecosystem studies tried to elucidate whether effects of a pesticide mixture can be explained by the effects of the single active substances when concentration or response addition is taken into account. Thus, there is a complete lack of data that could be used to assess the prediction quality of the two mixture toxicity concepts for the field situation. For this reason, this study has been initiated.

1.5 Aim of this study

In the context of the presented background information the following questions were addressed in this thesis:

Chapter 2

What role do reactive oxygen species play in copper toxicity to the two freshwater green algal species *Pseudokirchneriella subcapitata* and *Chlorella vulgaris*? Can intracellular concentrations of reactive oxygen species be linked to effects on photosynthesis?

Chapter 3

Does the concept of concentration addition hold true also for an algal community (phytoplankton) exposed to a mixture of the three PSII inhibitors atrazine, isoproturon, and diuron under field conditions? This was examined by investigating photosynthesis as an endpoint which is directly linked to the mode of action of the test substances.

Chapter 4

Is the concept of concentration addition also suitable to predict effects of a mixture of the three PSII inhibitors atrazine, isoproturon, and diuron on response parameters that describe higher levels of biological and ecological organization, such as abundance, diversity, and species composition of phytoplankton?

Chapter 5

Does a chronic exposure to environmentally relevant concentrations of the PSII inhibitors atrazine, diuron, and isoproturon and to a mixture of these herbicides lead to sustained effects on photosynthesis and growth of the three submersed aquatic macrophytes *Myriophyllum spicatum*, *Elodea canadensis*, and *Potamogeton lucens*?

1.6 References

Allan, R., 1997. Introduction: mining and metals in the environment. *Journal of Geochemical Exploration* 58, 95-100.

Altenburger, R., Backhaus, T., Boedeker, W., Faust, M., Scholze, M., Grimme, L.H., 2000. Predictability of the toxicity of multiple chemical mixtures to *Vibrio fischeri*: Mixtures composed of similarly acting chemicals. *Environmental Toxicology and Chemistry* 19, 2341-2347.

Andersson, B., Styring, S., 1991. Photosystem II - Molecular organization, function, and acclimation. *Current Topics in Bioenergetics* 16, 1-81.

Arrhenius, A., Gronvall, F., Scholze, M., Backhaus, T., Blanck, H., 2004. Predictability of the mixture toxicity of 12 similarly acting congeneric inhibitors of photosystem II in marine periphyton and epipsammon communities. *Aquatic Toxicology* 68, 351-367.

Arrhenius, A., Backhaus, T., Gronvall, F., Junghans, M., Scholze, M., Blanck, H., 2006. Effects of three antifouling agents on algal communities and algal reproduction: Mixture toxicity studies with TBT, Irgarol, and Sea-Nine. *Archives of Environmental Contamination and Toxicology* 50, 335-345.

Backhaus, T., Altenburger, R., Boedeker, W., Faust, M., Scholze, M., Grimme, L.H., 2000. Predictability of the toxicity of a multiple mixture of dissimilarly acting chemicals to *Vibrio fischeri*. *Environmental Toxicology and Chemistry* 19, 2348-2356.

Backhaus, T., Faust, M., Scholze, M., Gramatica, P., Vighi, M., Grimme, L.H., 2004a. Joint algal toxicity of phenylurea herbicides is equally predictable by concentration addition and independent action. *Environmental Toxicology and Chemistry* 23, 258-264.

Backhaus, T., Arrhenius, A., Blanck, H., 2004b. Toxicity of a mixture of dissimilarly acting substances to natural algal communities: Predictive power and limitations of independent action and concentration addition. *Environmental Science & Technology* 38, 6363-6370.

Bengtson Nash, S.M., Schreiber, U., Ralph, P.J., Müller, J.F., 2005. The combined SPE:ToxY-PAM phytotoxicity assay; application and appraisal of a novel biomonitoring tool for the aquatic environment. *Biosensors & Bioelectronics* 20, 1443-1451.

Berenbaum, M. C., 1985. The expected effect of a combination of agents: the general solution. *Journal of Theoretical Biology* 114, 413-431.

Bliss, C.I., 1939. The toxicity of poisons applied jointly. *Annals of Applied Biology* 26, 585-615.

Buchanan, B.B., Gruissem, W., Jones, R.L., 2000. *Biochemistry and Molecular Biology of Plants*, American Society of Plant Physiologists, Rockeville, MD, USA.

Calabrese, E.J., 1991. *Multiple Chemical Interactions*. Lewis Publishers, Chelsea, MI, USA.

Calamari, D., Alabaster, J.S., 1980. An approach to theoretical models in evaluating the effects of mixtures of toxicants in the aquatic environment. *Chemosphere* 9, 533-538.

Carter, A.D., 2000. Herbicide movement in soils: principles, pathways and processes. *Weed Research* 40, 113-122.

Cedeno-Maldonado, A., Swader, J. A., Heath, R. L., 1972. The cupric ion as an inhibitor of photosynthetic electron transport in isolated chloroplasts. *Plant Physiology* 50, 698-701.

Chèvre, N., Loeppe, C., Singer, H., Stamm, C., Fenner, K., Escher, B.I., 2006. Including mixtures in the determination of water quality criteria for herbicides in surface water. *Environmental Science & Technology* 40, 426-435.

Conrad, R., Buchel, C., Wilhelm, C., Arsalane, W., Berkaloff, C., Duval, J.C., 1993. Changes in yield in in vivo fluorescence of chlorophyll a as a tool for selective herbicide monitoring. *Journal of Applied Phycology* 5, 505-516.

Crommentuijn, T., Sijm, D., de Bruijn, J., van Leeuwen, K., van de Plassche, E., 2000. Maximum permissible and negligible concentrations for some organic substances and pesticides. *Journal of Environmental Management* 58, 297-312.

Devi, S.R., Prasad, M.N.V., 2005. Antioxidant capacity of *Brassica juncea* plants exposed to elevated levels of copper. *Russian Journal of Plant Physiology* 52, 205-8.

Dorigo, U., Leboulanger, C., 2001. A pulse-amplitude modulated fluorescence-based method for assessing the effects of photosystem II herbicides on freshwater periphyton. *Journal of Applied Phycology* 13, 509-515.

Drost, W., Backhaus, T., Vassilakaki, M., Grimme, L.H., 2003. Mixture toxicity of s-triazines to Lemna minor under conditions of simultaneous and sequential exposure. *Fresenius Environmental Bulletin* 12, 601-607.

EIFAC (European Inland Fisheries Advisory Committee), 1987. Water quality criteria for European freshwater fish. Revised report on combined effects on freshwater fish and other aquatic life of mixtures of toxicants in water. EIFAC Tech Pap 37, Rev 1.

ElJay, A., Ducruet, J.M., Duval, J.C., Pelletier, J.P., 1997. A high-sensitivity chlorophyll fluorescence assay for monitoring herbicide inhibition of photosystem II in the chlorophyte

Selenastrum capricornutum: Comparison with effect on cell growth. *Archiv für Hydrobiologie* 140, 273-286.

Fai, P.B., Grant, A., Reid, B., 2007. Chlorophyll a fluorescence as a biomarker for rapid toxicity assessment. *Environmental Toxicology and Chemistry* 26, 1520-1531.

Fairchild, J.F., Lapoint, T.W., Schwartz, T.R., 1994. Effects of an herbicide and insecticide mixture in aquatic mesocosms. *Archives of Environmental Contamination and Toxicology* 27, 527-533.

Falkowski, P.G., Raven, J.A., 2007. *Aquatic photosynthesis*. 2nd edition, Princeton University Press, Princeton, NJ, USA.

Faust, M., Altenburger, R., Backhaus, T., Blanck, H., Boedeker, W., Gramatica, P., Hamer, V., Scholze, M., Vighi, M., Grimme, L.H., 2001. Predicting the joint algal toxicity of multi-component s-triazine mixtures at low-effect concentrations of individual toxicants. *Aquatic Toxicology* 56, 13-32.

Faust, M., Altenburger, R., Backhaus, T., Blanck, H., Boedeker, W., Gramatica, P., Hamer, V., Scholze, M., Vighi, M., Grimme, L.H., 2003. Joint algal toxicity of 16 dissimilarly acting chemicals is predictable by the concept of independent action. *Aquatic Toxicology* 63, 43-63.

Fernandes, J.C., Henriques, F.S., 1991. Biochemical, physiological and structural effects of excess copper in plants. *Botanical Review* 57, 246-73.

Field, J.A., Reed, R.L., Sawyer, T.E., Griffith, S.M., Wigington, P.J., 2003. Diuron occurrence and distribution in soil and surface and ground water associated with grass seed production. *Journal of Environmental Quality* 32, 171-179.

Frei, W., 1913. Versuche über Kombinationen von Desinfektionsmitteln. *Zeitschrift für Hygiene Infektionskrankheiten* 75, 433-496.

Fuerst, E.P., Norman, M.A., 1991. Interactions of herbicides with photosynthetic electron transport. *Weed Science* 39, 458-464.

Graymore, M., Stagnitti, F., Allinson, G., 2001. Impacts of atrazine in aquatic ecosystems. *Environment International* 26, 483-495.

Hansson, O., Wydrzynski, T., 1990. Current perceptions of photosystem II. *Photosynthesis Research* 23, 131-162.

Hartgers, E.M., Aalderink, G.H.R., Van den Brink, P.J., Gylstra, R., Wiegman, J.W.F., Brock, T.C.M., 1998. Ecotoxicological treshold levels of a mixture of herbicides (atrazine, diuron and metolachlor) in freshwater microcosms. *Aquatic Ecology* 32, 135-152.

Hewlett, P.S., Plackett, R.L., 1952. Similar Joint Action of Insecticides. *Nature* 169, 198-199.

Huber, W., 1993. Ecotoxicological Relevance of Atrazine in Aquatic Systems. *Environmental Toxicology and Chemistry* 12, 1865-1881.

Irace-Guigand, S., Aaron, J.J., Scribe, P., Barcelo, D., 2004. A comparison of the environmental impact of pesticide multiresidues and their occurrence in river waters surveyed by liquid chromatography coupled in tandem with UV diode array detection and mass spectrometry. *Chemosphere* 55, 973-981.

Jahnel, J., Zwiener, C., Gremm, T.J., Abbt-Braun, G., Frimmel, F.H., Kussatz, C., Schudoma, D., Rocker, W., 2001. Quality targets for pesticides and other pollutants in surface waters. *Acta Hydrochimica et Hydrobiologica* 29, 246-253.

Johnson, A.K.L., Ebert, S.P., 2000. Quantifying inputs of pesticides to the Great Barrier Reef Marine Park - A case study in the Herbert river catchment of north-east Queensland. *Marine Pollution Bulletin* 41, 302-309.

Junghans, M., Backhaus, T., Faust, M., Scholze, M., Grimme, L.H., 2003a. Predictability of combined effects of eight chloroacetanilide herbicides on algal reproduction. *Pest Management Science* 59, 1101-1110.

Junghans, M., Backhaus, T., Faust, M., Scholze, M., Grimme, L.H., 2003b. Toxicity of sulfonylurea herbicides to the green alga *Scenedesmus vacuolatus*: Predictability of combination effects. *Bulletin of Environmental Contamination and Toxicology* 71, 585-593.

Knauer, K., Behra, R., Sigg, L., 1997. Effects of free Cu^{2+} and Zn^{2+} ions on growth and metal accumulation in freshwater algae. *Environmental Toxicology and Chemistry* 16, 220-229.

Koblizek, M., Masojidek, J., Komenda, J., Kucera, T., Pilloton, R., Mattoo, A.K., Giardi, M.T., 1998. A sensitive photosystem II-based biosensor for detection of a class of herbicides. *Biotechnology and Bioengineering* 60, 664-669.

Kotrikla, A., Gatidou, G., Lekkas, T.D., 2006. Monitoring of triazine and phenylurea herbicides in the surface waters of Greece. *Journal of Environmental Science and Health Part B-Pesticides Food Contaminants and Agricultural Wastes* 41, 135-144.

Krause, G.H., Weis, E., 1991. Chlorophyll fluorescence and photosynthesis - the basics. *Annual Review of Plant Physiology and Plant Molecular Biology* 42, 313-349.

Kreuger, J., 1998. Pesticides in stream water within an agricultural catchment in southern Sweden, 1990-1996. *Science of the Total Environment* 216, 227-251.

Leu, C., Singer, H., Stamm, C., Muller, S.R., Schwarzenbach, R.P., 2004. Variability of herbicide losses from 13 fields to surface water within a small catchment after a controlled herbicide application. *Environmental Science & Technology* 38, 3835-3841.

Loewe, S., Muischnek, H., 1926. Über die Kombinationswirkungen. 1. Mitteilung: Hilfsmittel der Fragestellung. *Naunyn Schmiedebergs Archiv der experimentellen pathologischen Pharmakologie* 114, 313-326.

Macfarlane, G.R., Burchett, M.D., 2001. Photosynthetic pigments and peroxidase activity as indicators of heavy metal stress in the grey mangrove, *Avicennia marina* (Forsk.) Vierh. *Marine Pollution Bulletin* 42, 233-240.

Mackay D, Shiu W.Y., Ma K., Lee S.C., 2006. *Handbook of physical-chemical properties and environmental fate for organic chemicals*. CRC Press, Taylor & Francis Group, Boca Raton, FL, USA.

Mallick, N., 2004. Copper-induced oxidative stress in the chlorophycean microalga *Chlorella vulgaris*: response of the antioxidant system. *Journal of Plant Physiology* 161, 591-597.

Merlin, G., 1997. Herbicides, in: Plant Ecophysiology, ed. Prasad, M.N.V., John Wiley & Sons, New York, USA.

Michel, H., Epp, O., Deisenhofer, J., 1986. Pigment-Protein Interactions in the Photosynthetic Reaction Center from Rhodopseudomonas-Viridis. *EMBO (European Molecular Biology Organization) Journal* 5, 2445-2452.

Müller, K., Bach, M., Hartmann, H., Spiteller, M., Frede, H.G., 2002. Point- and nonpoint-source pesticide contamination in the Zwester Ohm catchment, Germany. *Journal of Environmental Quality* 31, 309-318.

Nanba, O., Satoh, K., 1987. Isolation of a photosystem-II reaction center consisting of D1 and D2 polypeptides and cytochrome B559. *Proceedings of the National Academy of Sciences of the United States of America* 84, 109-112.

Nitschke, L., Schussler, W., 1998. Surface water pollution by herbicides from effluents of waste water treatment plants. *Chemosphere* 36, 35-41.

Pätsikkä, E., Aro, E. M., Tyystjärvi, E., 2001. Mechanism of copper-enhanced photoinhibition in thylakoid membranes. *Physiologia Plantarum* 113, 142-150.

Reynolds, C.S., 2006. *The Ecology of Phytoplankton*. Cambridge University Press, Cambridge, UK.

Scholz, N.L., Truelove, N.K., Labenia, J.S., Baldwin, D.H., Collier, T.K., 2006. Dose-additive inhibition of chinook salmon acetylcholinesterase activity by mixtures of organophosphate and carbamate insecticides. *Environmental Toxicology and Chemistry* 25, 1200-1207.

Schreiber, U., Bilger, W., Neubauer, C., 1994. *Chlorophyll Fluorescence as a Nonintrusive Indicator for Rapid Assessment of In Vivo Photosynthesis*, in: Ecophysiology of Photosynthesis, eds. Schulze, E.D, Caldwell, M.M., Springer Verlag, New York, USA, pp. 49-70.

Schwarzenbach, R.P., Escher, B.I., Fenner, K., Hofstetter, T.B., Johnson, C.A., von Gunten, U., Wehrli, B., 2006. The challenge of micropollutants in aquatic systems. *Science* 313, 1072-1077.

Silva, E., Rajapakse, N., Kortenkamp, A., 2002. Something from "nothing" - Eight weak estrogenic chemicals combined at concentrations below NOECs produce significant mixture effects. *Environmental Science & Technology* 36, 1751-1756.

Snel, J.F.H., Vos, J.H., Gylstra, R., Brock, T.C.M., 1998. Inhibition of Photosystem II (PSII) electron transport as a convenient endpoint to assess stress of the herbicide linuron on freshwater plants. *Aquatic Ecology* 32, 113-123.

Solomon, K.R., Baker, D.B., Richards, R.P., Dixon, D.R., Klaine, S.J., LaPoint, T.W., Kendall, R.J., Weisskopf, C.P., Giddings, J.M., Giesy, J.P., Hall, L.W., Williams, W.M., 1996. Ecological risk assessment of atrazine in North American surface waters. *Environmental Toxicology and Chemistry* 15, 31-74.

Sørensen, S.R., Ronen, Z., Aamand, J., 2001. Isolation from agricultural soil and characterization of a *Sphingomonas* sp able to mineralize the phenylurea herbicide isoproturon. *Applied and Environmental Microbiology* 67, 5403-5409.

Sprague, J.B., 1970. Measurement of pollutant toxicity to fish. II. Utilizing and applying bioassay results. *Water Research* 4, 3-32.

Stephan, C., Mount, D., Hansen, D., Gentile, J., Chapman, G., Brungs, W., 1985. *Guidelines for deriving numerical national water quality criteria for the protection of aquatic organisms and their uses*; NTIS no PB85-227049; US Environmental Protection Agency, Washington DC.

Streibig, J.C., Kudsk, P., 1993. *Herbicide Bioassays*. CSR Press, Boca Raton, FL, USA.

Sunda, W. G. 1988. Trace metal interaction with marine phytoplankton. *Biological Oceanography* 6, 411-441.

Tomlin, C.D.S., 2006. *The Pesticide Manual*. 14th edition, BCPC Publications, Hampshire, UK.

Trebst, A., 1987. The 3-dimensional structure of the herbicide binding niche on the reaction center polypeptides of photosystem II. *Zeitschrift für Naturforschung C - Journal of Biosciences* 42, 742-750.

Van den Brink, P.J., Hartgers, E.M., Gylstra, R., Bransen, F., Brock, T.C.M., 2002. Effects of a mixture of two insecticides in freshwater microcosms: II. Responses of plankton and ecological risk assessment. *Ecotoxicology* 11, 181-197.

Vighi, M., Altenburger, R., Arrhenius, A., Backhaus, T., Bodeker, W., Blanck, H., Consolaro, F., Faust, M., Finizio, A., Froehner, K., Gramatica, P., Grimme, L.H., Gronvall, F., Hamer, V., Scholze, M., Walter, H., 2003. Water quality objectives for mixtures of toxic chemicals: problems and perspectives. *Ecotoxicology and Environmental Safety* 54, 139-150.

Vroumsia, T., Steiman, R., SeigleMurandi, F., BenoitGuyod, J.L., Khadrani, A., 1996. Biodegradation of three substituted phenylurea herbicides (chlorotoluron, diuron, and isoproturon) by soil fungi. A comparative study. *Chemosphere* 33, 2045-2056.

Wauchope, R.D., 1978. Pesticide content of surface water draining from agricultural fields - Review. *Journal of Environmental Quality* 7, 459-472.

Wendt-Rasch, L., Pirzadeh, P., Woin, P., 2003. Effects of metsulfuron methyl and cypermethrin exposure on freshwater model ecosystems. *Aquatic Toxicology* 63, 243-256.

Wendt-Rasch, L., Van den Brink, P.J., Crum, S.J.H., Woin, P., 2004. The effects of a pesticide mixture on aquatic ecosystems differing in trophic status: responses of the macrophyte *Myriophyllum spicatum* and the periphytic algal community. *Ecotoxicology and Environmental Safety* 57, 383-398.

Wetzel, R.G., 2001. *Limnology - Lake and River ecosystems*. 3rd edition, Academic Press, San Diego, CA, USA.

Yruela, I., Pueyo, J. J., Alonso, P. J., Picorel, R., 1996a. Photoinhibition of photosystem II from higher plants - Effect of copper inhibition. *Journal of Biological Chemistry* 271, 27408-27415.

Zabel, T. and Cole, S.M., 1999. The derivation of environmental quality standards for the protection of aquatic life in the UK. *The Chartered Institution of Water and Environmental Management Journal* 13, 436-440.

Chapter 2

The role of reactive oxygen species in copper toxicity to two freshwater green algae

2.1 Abstract

The role of reactive oxygen species (ROS) in copper (Cu) toxicity to two freshwater green algal species, *Pseudokirchneriella subcapitata* and *Chlorella vulgaris*, was assessed to gain a better mechanistic understanding of Cu toxicity. Cu-induced formation of ROS was investigated in the two algal species and linked to short-term effects on photosynthetic activity and to long-term effects on cell growth. A light- and time-dependent increase in ROS concentrations was determined upon exposure to environmentally relevant Cu concentrations of 50 nM and 250 nM and was comparable in both algal species. However, effects of 250 nM Cu on photosynthesis were different leading to a 12 % reduction in photosynthetic activity in *P. subcapitata* but not in *C. vulgaris*. These results indicate that differences in species-specific sensitivities measured as photosynthetic activity were not caused by differences in the cellular ROS content of the algae, but probably by different species-specific ROS defense systems. To investigate the role of ROS in Cu mediated inhibition of photosynthesis, the ROS scavenger N-tert-butyl-α-phenylnitrone (BPN) was used, resulting in a reduction of Cu-induced ROS production up to control level and a complete restoration of photosynthetic activity of Cu-exposed *P. subcapitata*. This finding implied that ROS play a primary role in Cu toxicity to algae. Furthermore, we observed a time-dependent ROS release process across the plasma membrane. More than 90 % of total ROS were determined to be extracellular in *P. subcapitata*, indicating an efficient way of cellular protection against oxidative stress.

2.2 Introduction

Copper (Cu) is an essential micronutrient for phototrophic organisms, such as cyanobacteria, algae, and higher plants. It is an important constituent of several metalloenzymes that are mostly involved in the catalysis of redox actions with O_2 as electron acceptor and of superoxide dismutase, an important enzyme in antioxidant defense mechanism. In addition, Cu was found as an indispensable compound in some parts of the photosynthetic apparatus, namely in the plastocyanin of PSI and in the light-harvesting complex of PSII (Fernandes and Henriques 1991). Availability of metal ions to algae, as well as toxic effects, strongly depend on the chemical speciation (Sunda

1988). In many studies, biological effects have been shown to be related to the free metal ion concentration (Sunda and Guillard 1976, Morel et al. 1978, Knauer et al. 1997). Elevated Cu concentrations are highly toxic to algae and higher plant tissues. The toxicity of Cu cannot be related to one specific mode of action. It interferes with several aspects of plant biochemistry, including photosynthesis, chlorophyll synthesis, fatty acid metabolism and carbohydrate synthesis and is therefore an effective inhibitor of vegetative growth (Fernandes and Henriques 1991). Cu is known for its strong binding affinity to sulfhydryl groups, which are essential for enzymatic activity and protein structure (De Filippis and Pallaghy 1994). The most important effect of Cu on phototrophic organisms is associated with the inhibition of photosynthesis. The sensitivity of the photosynthetic apparatus to excess copper was first demonstrated by Macdowell (1949). Further investigations found PSII to be more sensitive to Cu inhibition than PSI (Cedeno-Maldonado et al. 1972). PSII is a large protein complex - embedded in the thylakoid membrane of the chloroplasts and consisting of about 20 protein subunits - that catalyzes the light-driven reduction of plastoquinone by electrons from water that is oxidized to molecular oxygen (Andersson and Styring 1991, Vermaas et al. 1993). The mechanism of Cu toxicity to photosynthetic electron transport has been widely investigated in a number of *in vitro* and *in vivo* studies. Nevertheless, the precise location of the Cu inhibitory binding site still remains a matter of controversy. On the one hand, the target site of Cu inhibition of PSII is thought to be located at its acceptor side (Mohanty et al. 1989, Yruela et al. 1991, 1993). Cu interactions with the pheophytin-Q_A-Fe^{2+} domain or Cu-induced modifications in the amino acid or lipid structure close to the Q_A and Q_B binding sites have been suggested to cause the blockage of photosynthetic electron transport (Yruela et al. 1996a,b). In contrast, other studies conclude that Cu impaired PSII electron transport on the donor side by an interaction at or beyond the PSII primary electron carrier donor, Tyr161, the redox active tyrosine residue of the D1 protein (Schröder et al. 1994, Pätsikkä et al. 2001). Furthermore, it was demonstrated that the central magnesium atom of chlorophyll can be substituted by several metals (e.g., mercury, Cu or cadmium), damaging the photosystem (Küpper et al. 1996). Nevertheless, all investigations on the specific Cu inhibitory binding site imply direct interferences of the metal ion with the photosynthetic apparatus, resulting in a reduced electron flow.

Besides the controversial discussion about the precise target site of copper inhibition, various authors gave indirect evidence that the toxic effect of Cu to phototrophic organisms strongly appears to be related to the production of reactive oxygen species (ROS) since they observed a Cu-induced activation of the antioxidant defence system as well as an increase in the levels of lipid peroxidation and protein carbonylation (Gallego et al. 1996, Rama Devi and Prasad 1998, Nagalakshmi and Prasad 1998, Okamoto and Colepicolo 1998, Collén et al. 2003, Ratkevicius et al.

2003, Mallick 2004, Devi and Prasad 2005, Tripathi et al. 2006). Oxidative damage induced by Cu is thought to be a consequence, rather than a direct cause, of Cu toxicity (Halliwell and Gutteridge 1999).

Organisms have evolved numerous protective mechanisms that serve to scavenge ROS before they can severely damage sensitive parts of the cellular machinery. Defence against ROS includes low molecular mass agents such as glutathione, ascorbate, flavonoids, α-tocopherol, and carotenoids as well as enzymatic catalysts of high molecular weight (e.g., catalase, superoxide dismutase, peroxidases) (Imlay 2003, Pinto et al. 2003).

In the present study, we investigated ROS formation upon Cu exposure under light and dark conditions and linked cellular ROS levels to effects on photosynthesis of the two freshwater green algal species *P. subcapitata* and *C. vulgaris*. Direct measurements of total ROS were made using a fluorometric approach. We determined whether ROS are merely a consequence of copper toxicity or if ROS play a primary role in Cu toxicity towards algal cells by applying a ROS scavenging agent. Furthermore, we studied a possible protection and detoxification mechanism against oxidative stress that involves the export of ROS across the plasma membrane.

2.3 Materials and methods

2.3.1 Cu exposure

The nominal Cu concentrations were 50 nM and 250 nM in the experiments. These initial concentrations were analytically verified by inductively coupled plasma – atomic emission spectrophotometry (Spectro, Kleve, Germany) resulting in 52 ± 10 nM (50 nM) and 262 ± 19 nM (250 nM). Cu concentration in the OECD medium (Organisation for Economic Co-operation and Development, OECD Guideline 201), which served as control, was 26 ± 14 nM due to background contamination of the applied salts, and thus the control was not Cu deficient. The nominal Cu concentrations of 50 nM and 250 nM are used herein to describe the Cu treatments.

The concentrations of the free Cu^{2+} ion in the medium were computed with the chemical speciation program ChemEQL 3.0 (Beat Müller, eawag, Kastanienbaum, Switzerland) resulting in 1.9×10^{-11} M, 4.2×10^{-11} M, and 3.3×10^{-10} M for the control, and the 50 and 250 nM Cu treatments, respectively.

2.3.2 Test organisms and culture conditions

The two freshwater green algal species, *Pseudokirchneriella subcapitata* (Korshikov) Hindák (SAG 61.81) and *Chlorella vulgaris* Beij. (SAG 211-11b), were purchased as stock cultures on agar from the Culture Collection of Algae (SAG, University of Göttingen, Germany). A liquid stock culture of the two strains was grown in 10-fold concentrated OECD algal test medium with slight modifications. $CuCl_2·2(H_2O)$ was not added. To keep the pH constant at 7.4 during the experiments, the medium was supplemented with 10 mM Hepes. $NaNO_3$ (3 mM) instead of NH_4Cl served as the nitrogen source. All salts used for the preparation of the algal medium were obtained from Sigma-Aldrich (Buchs, Switzerland). The algal cells were grown at 100 rpm and 25 °C in a Multitron incubator (Infors AG, Basel-Bottmingen, Switzerland). An illumination of 40 µmol m^{-2} s^{-1} was provided by SYLVANIA GRO-LUX fluorescent light (F15W/Gro T8; Infors AG) under a light:dark regime of 16 h:8 h.

All experiments were conducted with exponentially growing cells. Algal test cultures were inoculated with 10^5 cells·mL^{-1} and grown for 4 to 6 days until they reached a cell density ranging between 1 x 10^6 and 3 x 10^6 cells·mL^{-1}. A linear correlation fit between cell density determined using a Neubauer counting chamber and the optical density of the two algal species determined with a Helios α 4.0 spectrophotometer (Thermo Spectronics, Cambridge, UK) was prepared. Cell density was assessed by measuring the optical density of *P. subcapitata* and *C. vulgaris* cell suspensions at 680 nm and 684 nm. For all experiments described below, the test medium was modified OECD medium.

2.3.3 Detection of ROS formation

ROS formation was measured by using the cell permeable indicator 2′,7′dichlorodihydrofluorescein diacetate (H_2DCFDA, Molecular Probes, Eugene, OR, USA). Cellular esterases hydrolyze the probe to the nonfluorescent 2′,7′dichlorodihydrofluorescein (H_2DCF) which is better retained in the cells. In the presence of ROS and cellular peroxidases, H_2DCF is transformed to the highly fluorescent 2′,7′dichlorofluorescein (DCF) (Haugland 2005) (Fig. 2.1). The stock solution of H_2DCFDA was prepared in methanol at a concentration of 10 mM and stored at -80 °C.

For the determination of ROS, algal cells of *P. subcapitata* and *C. vulgaris* were centrifuged at 2000 g for 5 min, and the cell pellet was resuspended in OECD medium. Subsequently, several wells of a 96-well microplate received 1 x 10^7 (*P. subcapitata*) or 2.5 x 10^7 (*C. vulgaris*) algal cells, followed immediately by the addition of 50 or 250 nM Cu and 5 µM H_2DCFDA. Control wells received only algal cells and H_2DCFDA in OECD medium. Solvent controls were performed similarly. Methanol concentrations did not exceed 0.05 % in solvent control and treated wells. The final volume was 200 µL per well. Fluorescence of DCF was measured every 30 min up to 270 min with a Chameleon multilabel microplate reader (Hidex, Turku, Finland) set at a photomultiplier gain of 30, with excitation and emission filters of 485 nm and 530 nm, respectively. The total ROS formation was determined under light and dark conditions. As the fluorescent probe and its oxidation product are light sensitive substances, the experiments under light conditions were performed under an illumination of 15 µE m^{-2} s^{-1}. DCF fluorescence data were expressed as absolute fluorescent units (FU). DCF formation rates (FU DCF/min) in relation to control were determined by linear regression for the linear period of fluorescence increase over time. Conversion of H$_2$DCFDA to the fluorescent product DCF induced by light in the absence of algal cells was subtracted as background level from the FU.

Total ROS concentrations consist of intra- and extracellular ROS (Fig. 2.1). To differentiate between intracellular ROS concentrations and the excretion of ROS, *P. subcapitata* was exposed to 50 and 250 nM total added Cu for different times. Subsequent to Cu incubation, the medium of replicated wells was removed, and Cu and dye-free medium was added to the algal cells. DCF fluorescence was determined prior to (DCF$_{total}$) and after medium change (DCF$_{intra}$). The extracellular DCF concentration was calculated as DCF$_{extra}$ = DCF$_{total}$ − DCF$_{intra}$.

The cell permeable scavenger N-tert-butyl-α-phenylnitrone (BPN, Sigma-Aldrich, Buchs, Switzerland) was applied to trap ROS. A stock solution of 300 mM BPN was prepared in methanol. Algal cells of *P. subcapitata* were simultaneously exposed to H$_2$DCFDA, 250 nM Cu and 300 µM BPN. Methanol concentrations were identical in the solvent control and treated wells, reaching 1.5 %. ROS formation was detected and determined as described above.

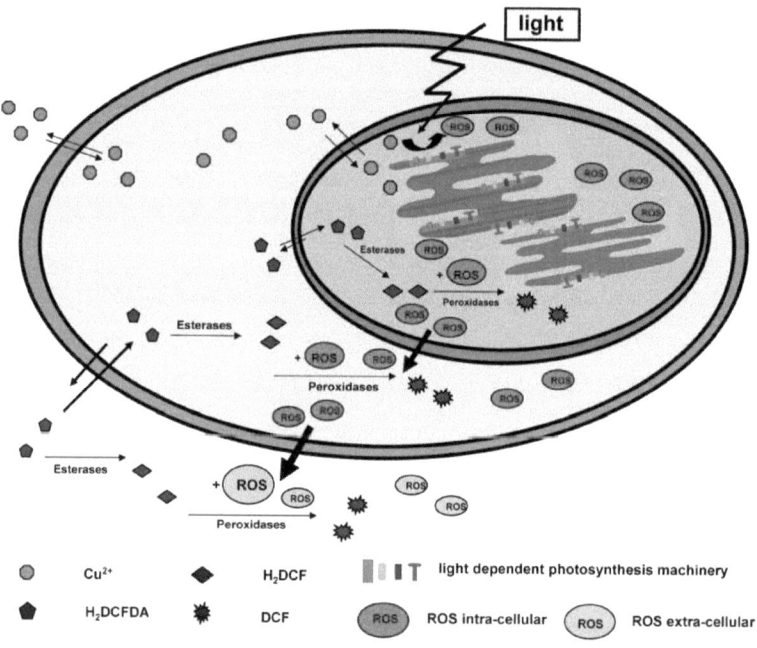

Fig. 2.1 This schematic view shows the transport and extra- and intracellular localization and conversion of the applied fluorogenic probe H$_2$DCFDA in an algal cell. Esterases deacetylate the probe into H$_2$DCF, which then can react, catalyzed by peroxidases, with ROS to the measurable fluorescence product DCF. Cu reaches its target site in the thylakoid membrane of the chloroplast leading to the generation of ROS. DCF, 2′,7′dichlorofluorescein; H$_2$DCF, 2′,7′dichloro-dihydrofluorescein; H$_2$DCFDA, 2′,7′dichlorodihydro-fluorescein diacetate; ROS, reactive oxygen species.

2.3.4 Determination of *in vivo* photosynthetic activity

Effects of Cu on the *in vivo* chlorophyll fluorescence of *P. subcapitata* and *C. vulgaris* were determined with the saturating pulse method using a pulse-amplitude modulation fluorometer (ToxY-PAM Dual Channel Yield Analyzer, Heinz Walz GmbH, Effeltrich, Germany) (Schreiber et al. 2002). The *in vivo* chlorophyll fluorescence measured as PSII quantum yield (Y) reflects photosynthetic activity of the algal cells and can be calculated according to the following equation:

$$Y = (F_m' - F)/F_m'$$

where F is the present fluorescence measured prior to the application of the saturating pulse and F_m' is the maximum fluorescence yield induced by the saturation pulse (Genty et al. 1989).

The *in vivo* photosynthetic activity of *P. subcapitata* and *C. vulgaris* was determined for cell suspensions with a density of 8 × 10^4 cells·mL^{-1} and 3 × 10^5 cells·mL^{-1}, respectively, after 4 h and

24 h of Cu exposure. Different cell numbers for the two algal species were chosen to receive reasonable F_m' fluorescence values ranging between 2500 and 3400 (Schreiber 2001). Moreover, algal cells of *P. subcapitata* were simultaneously exposed to 250 nM Cu and an approximately 10-, 100- or 1000-fold higher concentration of BPN (3, 30, 300 µM) for 24 h. In BPN experiments, solvent concentrations reached 0.1 %. Corresponding solvent controls were also measured. The PS-II quantum yield was measured five times at 30 s intervals, whereas only the last three values were used for calculation of an average yield. Fluorescence data of the Cu treatments were expressed in relation to control values.

2.3.5 Determination of specific growth rates

A volume of 100 mL medium was inoculated with exponentially growing cells from the stock cultures of *P. subcapitata* and *C. vulgaris* to obtain an initial cell density of 1×10^5 cells·mL^{-1}. Initial Cu concentrations of 50 nM and 250 nM were added to the medium. Culture growth was determined at daily intervals for up to 4 days. The pH was kept constant at 7.4 during this time period. Specific growth rates (µ) of the two green algal species were determined in triplicates for both Cu treatments and the control according to the following equation:

$$\mu = (\ln N_t - \ln N_0)/t$$

whereas t is the time, N_t is the number of algal cells after a certain time interval t, and N_0 is the cell number at the beginning (t = 0).

2.4 Results

2.4.1 ROS formation

A time-dependent rise in DCF fluorescence as a measure of cellular ROS content was determined in both algal species under illuminated conditions (Fig. 2.2). Absolute ROS concentrations produced by the same number of cells were comparable in *C. vulgaris* and *P. subcapitata*. Under dark conditions, no significant increase in ROS production was observed upon exposure to 50 nM Cu and 250 nM Cu in both algal species (one way ANOVA followed by Dunnett's test; p > 0.05) (Fig. 2.3). The variability of ROS measurements, expressed as coefficient of variation (CV), was higher in *C. vulgaris* (CV_{max} = 43.5 %) compared to *P. subcapitata* (CV_{max} = 25.3 %). Thus, we decided to focus all further investigations on *P. subcapitata*.

To investigate the excretion of ROS by the algal cells, intra- and extracellular DCF fluorescence in the percentage of total DCF formation was monitored in *P. subcapitata* in a time-dependent manner (Fig. 2.4). Intra- and extracellular DCF values in the percentage of total DCF production were comparable for algal cells exposed to control medium and 250 nM Cu. Intracellular DCF corresponded to 86.5 ± 0.4 % (control) and 86.5 ± 0.6 % (250 nM Cu) of total DCF production 15 min after the addition of H_2DCFDA or H_2DCFDA and 250 nM Cu. Over time, intracellular DCF, as a percentage of total ROS, decreased, whereas extracellular DCF increased. After 150 min, the ratio of intra- and extracellular DCF reached a steady state. In control algal cells, the intra:extra DCF ratio was 5.9 ± 0.5 to 94.1 ± 2.1 % of total DCF. Algal cells exposed to 250 nM Cu had a similar intra:extra DCF ratio (4.2 ± 0.2 to 95.8 ±1.7 % of total DCF).

When algal cells of *P. subcapitata* were simultaneously exposed to 250 nM Cu and 300 µM BPN, ROS production was reduced up to that observed in the control treatment after 270 min (Fig. 2.5). The scavenger BPN (300 µM) alone also lowered ROS production in *P. subcapitata* under the control value.

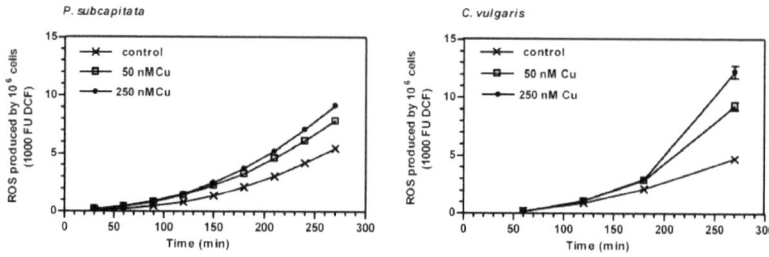

Fig. 2.2 Total reactive oxygen species (ROS) formation (measured and expressed as fluorescence units of dichlorofluorescein, FU DCF) in *P. subcapitata* and *C. vulgaris* upon exposure to 50 and 250 nM copper (Cu) over time under light conditions. Each data point represents mean ± standard error of five (*P. subcapitata*) or six (*C. vulgaris*) culture wells. Both diagrams are examples of four (*P. subcapitata*) and three (*C. vulgaris*) independent experiments.

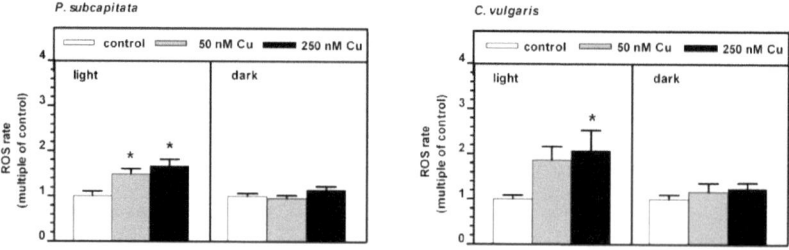

Fig. 2.3 Reactive oxygen species (ROS) formation rates (measured as increase in 2´,7´dichlorofluorescein fluorescence over the time) in *P. subcapitata* and *C. vulgaris* upon exposure to 50 and 250 nM Cu under light and dark conditions. ROS formation rates are expressed as a multiple of control. Each bar represents mean ± standard error of three to six experiments. An asterisk indicates a significant increased ROS rate compared to control (one-way ANOVA followed by Dunnett's test; $p<0.05$).

Chapter 2 – Copper induced oxidative stress in green algae

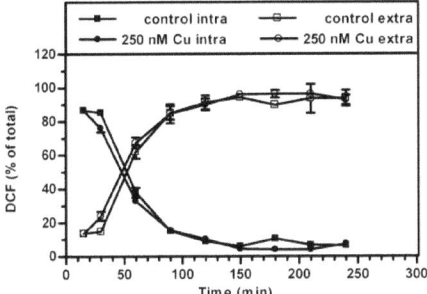

Fig. 2.4 Intra- and extracellular 2′,7′dichlorofluorescein (DCF) fluorescence in *P. subcapitata* expressed as % of total DCF fluorescence over time. Algal cells were exposed to control medium or 250 nM copper (Cu). Each data point represents mean ± standard error of three to four culture wells. The diagram shows an example of three independent experiments.

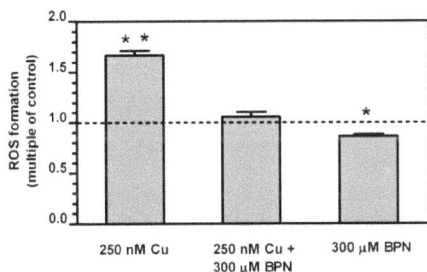

Fig. 2.5 Reactive oxygen species (ROS) formation (measured as 2′,7′dichlorofluorescein fluorescence) in *P. subcapitata* upon exposure to 250 nM copper (Cu), 250 nM Cu + 300 μM BPN and 300 μM BPN after 270 min. ROS formation is expressed as a multiple of control. Each bar represents mean ± standard error of five wells. One asterisk indicates a significantly decreased (two asterisks indicate increased) ROS formation compared to the control level (one-way ANOVA followed by Dunnett's test; $p<0.01$). BPN, N-tert-butyl-α-phenylnitrone.

2.4.2 *In vivo* photosynthetic activity

A concentration of 250 nM Cu decreased *in vivo* photosynthetic activity to 0.88 ± 0.02-fold relative to control (n = 4) after 24 h, whereas 50 nM Cu had no effect (Fig. 2.6). In *C. vulgaris*, neither 50 nM nor 250 nM Cu (Fig. 2.6) had a measurable effect, but concentrations ≥ 3 μM Cu (data not shown) impaired photosynthetic activity.

Incubation of *P. subcapitata* with 3, 30 or 300 μM BPN (Fig. 2.7) as well as with the solvent methanol (data not shown) showed no effects on photosynthetic activity after 24 h compared to control. A simultaneous exposure of *P. subcapitata* to 250 nM Cu and 3 µM BPN led to an inhibition of photosynthetic activity (0.88 ± 0.01-fold relative to control, n = 4), similar to that observed for 250 nM Cu alone. Algal cells that were exposed to 250 nM Cu in combination with 30 or 300 µM BPN exhibited a photosynthetic activity corresponding to control level.

In all experiments investigating ROS formation and photosynthetic activity, methanol controls and solvent-free medium controls were measured. In general, no significant difference between methanol, added at levels matching additions due to experimental treatments, and solvent-free medium control could be observed (data not shown).

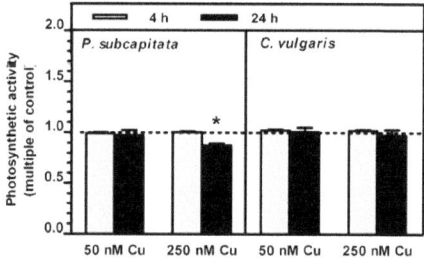

Fig. 2.6 *In vivo* photosynthetic activity (expressed as a multiple of control) was measured in *P. subcapitata* and *C. vulgaris* upon exposure to 50 and 250 nM copper (Cu) under light conditions after 4 and 24 h. Each bar represents mean ± standard error of three to five independent experiments. The asterisk indicates a significant decline in photosynthetic activity compared to the control level (one-way ANOVA followed by Dunnett's test; $p<0.05$).

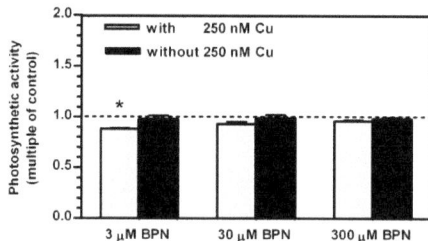

Fig. 2.7 *In vivo* photosynthetic activity (expressed as a multiple of control) was measured in *P. subcapitata* upon exposure to different concentrations of BPN and to 250 nM copper (Cu) in combination with different concentrations of BPN under light conditions after 24 h. Each bar represents mean ± standard error of three or four independent experiments. The asterisk indicates a significant decline in photosynthetic activity compared relative to control (one-way ANOVA followed by Dunnett's test; $p<0.05$). BPN, N-tert-butyl-α-phenylnitrone.

2.4.3 Specific growth rates

Cell growth of *P. subcapitata* and *C. vulgaris* was not impaired by the tested Cu concentrations, resulting in specific growth rates of 0.67 ± 0.05 d^{-1} (control), 0.72 ± 0.03 d^{-1} (50 nM), and 0.66 ± 0.02 d^{-1} (250 nM) for *P. subcapitata*, and 2.02 ± 0.10 d^{-1} (control), 1.76 ± 0.04 d^{-1} (50 nM) and 2.35 ± 0.14 d^{-1} (250 nM) for *C. vulgaris* (Fig. 2.8).

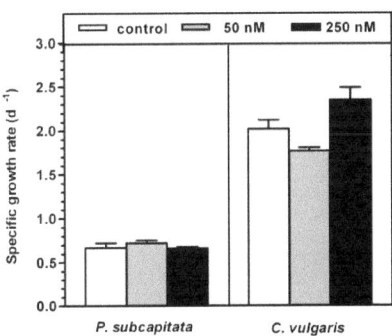

Fig. 2.8 Specific growth rates of *P. subcapitata* and *C. vulgaris* upon exposure to 50 and 250 nM copper (Cu). Bars represent mean ± standard error of three replicates. Treatments were not significantly different from control (one-way ANOVA followed by Dunnett's test; p>0.05).

2.5 Discussion

In this study, we demonstrated the formation of total ROS in algal cells exposed to environmentally relevant concentrations of 50 and 250 nM Cu by using the H_2DCFDA fluorometric approach. This method enabled the direct quantification of oxidative stress in the algal cells. In the context of Cu toxicity, this fluorometric assay has only been used to assess total ROS production in the human hepatoma cell line HepG2 (Seth et al. 2004), in trout hepatocytes (Manzl et al. 2004) and in the brown kelp *Lessonia nigrescens* (Andrade et al. 2006) so far. A direct determination of one specific group of ROS, namely superoxide radicals, was performed in Cu-exposed photosynthetic membranes of isolated spinach chloroplasts by Sandmann and Böger (1980). However, in most studies, indirect methods were applied to address Cu-induced oxidative stress in phototrophic organisms by determining changes in the antioxidant defence system or oxidative damage to proteins and lipids (Rijstenbil et al. 1994, Gallego et al. 1996, Okamoto et al. 1998, Collén et al. 2003, Mallick 2004, Morelli and Scarano 2004, Devi and Prasad 2005, Li et al. 2006). In contrast to our study, these studies investigated effects induced by higher total Cu concentrations ranging between 3 and 500 µM. In this study, the calculated free Cu^{2+} concentrations in the control was two to four orders of magnitude higher than concentrations in uncontaminated Swiss lakes (Xue and Sigg 1993), and the free Cu^{2+} concentrations in the 250 nM treatment was comparable to Cu^{2+} concentrations in the metal-contaminated Lake Orta, Italy (Knauer et al. 1998).

Our findings strongly indicate that Cu-stimulated ROS generation in *P. subcapitata* and *C. vulgaris* is a light dependent mechanism. The fact that light is required for the expression of the toxic effects of Cu was also demonstrated in isolated spinach chloroplasts (Cedeno-Maldonado et al. 1972, Uribe and Stark 1982), in *Chlorella pyrenoidosa* (Steemann-Nielsen et al. 1969), and in the cyanobacterium *Spirulina platensis* (Lu and Zhang 1999).

The ROS scavenging agent BPN completely restored photosynthetic activity of Cu exposed *P. subcapitata*. For this reason, Cu-induced oxidative damage to essential protein and lipid components, which are embedded in the thylakoid membrane and are therefore directly involved in photosynthesis, might be primarily responsible for the inhibition of electron flow. This would imply that the generation of ROS is the causal agent of Cu toxicity to algal cells. These findings are in contrast to the current opinion that oxidative damage is merely a consequence of Cu toxicity (Halliwell and Gutteridge 1999) and that a direct interaction of Cu with either the donor or with the acceptor side of PSII leads to the interruption of electron transport (Mohanty et al. 1989, Yruela et al. 1991, Schröder et al. 1994, Pätsikkä et al. 2001). We further assume that a ROS generating process involving a Cu-catalyzed Haber-Weiss or Fenton type reaction mechanism (Apel and Hirt 2004, Leonard et al. 2004) might also not play a major role in Cu toxicity to photosynthesis in algal cells, as it would reduce the electron flux in thylakoid membranes in the absence of ROS in a similar manner. Hence, other processes must be involved in the observed light dependent Cu stimulated ROS generation mechanism.

In this study, photosynthetic activity of *C. vulgaris* and *P. subcapitata* was affected differently by Cu. Whereas adverse effects on photosynthesis in *P. subcapitata* were already observed for 250 nM Cu, we noticed an inhibition of photosynthetic activity in *C. vulgaris* at Cu concentrations ≥ 3 µM (data not shown). For *Chlorella pyrenoidosa*, Steemann Nielsen et al. (1969) had results comparable to *P. subcapitata* by measuring photosynthesis rate as ^{14}C uptake under similar exposure conditions. The different effects of Cu on photosynthesis cannot be simply explained by the cellular ROS content since this was comparable between the two algal species tested in this study. Consequently, we assume that species-specific differences in the respective algal ROS defence systems are probably responsible for this observation. That the specific activities of antioxidative enzymes can be different upon exposure to 3 µM Cu was previously shown for the macrophyte *Ceratophyllum demersum* and the red macroalgae *Gracilaria tenuistipitata* (Rama Devi and Prasad 1998, Collén et al. 2003). In *C. demersum*, catalase appeared to be the most important antioxidative enzyme in ROS defence whereas in *G. tenuistipitata* the specific activity of ascorbate peroxidase was highest. On the contrary, 10 µM Cu did not impair ascorbate peroxidase activity at all in the marine diatom *Phaeodactylum tricornutum*.

Although short term effects of Cu on photosynthetic activity varied in *P. subcapitata* and *C. vulgaris*, no long term effects on algal cell growth were observed, as 50 and 250 nM Cu did not alter specific growth rates of both species compared to control cultures. Knauer et al. (1997) obtained comparable results, i.e., cell growth of the two green algal species *Scenedesmus subspicatus* and *Chlorella fusca* was optimal over a broad range of total Cu (10^{-9} to 10^{-3} M). It was further shown that algae have evolved several cellular adaptations to Cu toxicity providing long-term metal tolerance, for instance, the synthesis of metal chelating agents (e.g. metallothioneins, phytochelatins) (Knauer et al. 1998, Cobbett and Goldsbrough 2002) or active efflux of Cu ions by ATP driven pumps (Cumming and Gregory 1990).

In *P. subcapitata*, DCF fluorescence was predominantly measured in the surrounding medium of the algal cells after 150 min, which would indicate either leakage of intracellular-formed DCF or extracellular conversion of H_2DCFDA in the presence of released ROS (Fig. 2.1). We assume that ROS molecules generated in the chloroplasts diffused across the thylakoid and plasma membrane. Extracellular esterases and peroxidases subsequently converted H_2DCFDA to the fluorescent product DCF in the presence of released ROS. Extracellular peroxidase activity, for instance, was shown for the green alga *Selenastrum minutum* (Weger 1997). Simple leakage of DCF seemed to be unlikely since this molecule is polar and therefore more or less membrane impermeable. In addition, in the brown alga *Fucus evanescens*, DCF leakage was already shown to be minimal (Collén and Davison 1997). We suppose hydrogen peroxide (H_2O_2) to be the predominant ROS measured in the algal cells since it is a relatively stable (half life of 1 ms) (Reth 2002) and thus a less reactive ROS molecule compared with other active oxygen species, such as hydroxyl radical, superoxide anion radical, or singlet oxygen. Excretion of H_2O_2 has also been reported for other organisms, such as cyanobacteria (Stevens et al. 1973), green micro- and macroalgae (Zepp et al. 1987, Collén et al. 1995), and red macroalgae (Collén et al. 1994). In *Euglena gracilis*, Ishikawa et al. (1993) demonstrated that H_2O_2 is generated in chloroplasts and mitochondria and immediately diffuses from each organelle to the cytosol. The release of ROS can be an important process to protect the cells against enhanced cellular ROS concentrations, as previously reported for the filamentous green alga *Enteromorpha ahlneriana* under conditions of elevated oxidative stress induced by high irradiance (Choo et al. 2004). The discharge of intracellularly produced H_2O_2 appeared to be a facilitated diffusion process across membranes as we observed a constant intra:extra ROS ratio of approximately 1:9 of total ROS in Cu-exposed, as well as in nonexposed, algal cells. Since it was shown that biomembranes are poorly permeable to H_2O_2 (Antunes and Cadenas 2000, Seaver and Imlay 2001, Makino et al. 2004), we suggest that aquaporins may possibly play a role in the transport of H_2O_2 across the algal thylakoid and plasma membrane. Aquaporins were previously discussed in the context of a vacuolar ROS detoxification system in plants by Bienert et al. (2006). Besides various known protection mechanisms against ROS, including low molecular mass agents or antioxidant defence enzymes (Pinto et. al 2003), our results indicate that the release of ROS might be an efficient way to overcome oxidative stress.

2.6 References

Andersson, B., Styring, S., 1991. Photosystem II: Molecular organization, function, and acclimation. *Current Topics in Bioenergetics* 16, 1-81.

Andrade, S., Contreras, L., Moffett, J.W., Correa, J.A., 2006. Kinetics of copper accumulation in *Lessonia nigrescens* (Phaeophyceae) under conditions of environmental oxidative stress. *Aquatic Toxicology* 78, 398-401.

Antunes, F., Cadenas, E., 2000. Estimation of H_2O_2 gradients across biomembranes. *FEBS Letters* 475, 121-126.

Apel, K., Hirt, H., 2004. Reactive oxygen species: Metabolism, oxidative stress, and signal transduction. *Annual Review in Plant Biology* 55, 373-399.

Bienert, G.P., Schjoerring, J.K., Jahn, T.P., 2006. Membrane transport of hydrogen peroxide. *Biochimica et Biophysica Acta - Biomembranes* 1758, 994-1003.

Cedeno-Maldonado, A., Swader, J.A., Heath, R.L., 1972. The cupric ion as an inhibitor of photosynthetic electron transport in isolated chloroplasts. *Plant Physiology* 50, 698-701.

Choo, K.S., Snoeijs, P., Pedersen, M., 2004. Oxidative stress tolerance in the filamentous green algae *Cladophora glomerata* and *Enteromorpha ahlneriana*. *Journal of Experimental Marine Biology and Ecology* 298, 111-123.

Cobbett, C., Goldsbrough, P., 2002. Phytochelatins and metallothioneins: Roles in heavy metal detoxification and homeostasis. *Annual Review in Plant Biology* 53, 159-182.

Collén, J., Ekdahl, A., Abrahamsson, K., Pedersén, M., 1994. The involvement of hydrogen peroxide in the production of volatile halogenated compounds by *Meristiella gelidium*. *Phytochemistry* 36, 1197-1202.

Collén, J., Delrio, M. J., Garciareina, G., Pedersen, M., 1995. Photosynthetic production of hydrogen peroxide by *Ulva rigida* C. Ag. (Chlorophyta). *Planta* 196, 225-230.

Collén, J., Davison, I.R., 1997. In vivo measurement of active oxygen production in the brown alga *Fucus evanescens* using 2',7'-dichlorohydrofluorescein diacetate. *Journal of Phycology* 33, 643-648.

Collén, J., Pinto, E., Pedersen, M., Colepicolo, P., 2003. Induction of oxidative stress in the red macroalga *Gracilaria tenuistipitata* by pollutant metals. *Archives of Environmental Contamination and Toxicology* 45, 337-42.

Cumming, J.R., Gegory, J.T., 1990. *Mechanism of metal tolerance in plants: physiological adaptations for exclusion of metal ions from cytoplasm*, in: Stress Responses in Plants: Adaptation and Acclimatation Mechanisms, eds. Alscher, R.G., Cumming, J.R., Wiley-Liss, New York, NY, USA, pp. 338-55.

De Filippis, L.F., Pallaghy, C.K., 1994. *Heavy metals: sources and biological effects*, in: Algae and water pollution, eds. Rai, L.C., Gaur, J.P., Soeder, C.J., Schweizerbart'sche Verlagsbuchhandlung, Stuttgart, Germany, pp. 31-37.

Devi, S.R., Prasad, M.N.V., 2005. Antioxidant capacity of Brassica juncea plants exposed to elevated levels of copper. *Russian Journal of Plant Physiology* 52, 205-8.

Fernandes, J.C., Henriques, F.S., 1991. Biochemical, physiological and structural effects of excess copper in plants. *Botanical Review* 57, 246-73.

Gallego, S.M., Benavides, M.P., Tomaro, M.L., 1996. Effect of heavy metal ion excess on sunflower leaves: evidence for involvement of oxidative stress. *Plant Science* 121, 151-9.

Genty, B., Briantais, J. M., Baker, N., 1989. The relationship between the quantum yield of photosynthetic electron transport and quenching of chlorophyll fluorescence. *Biochimica et Biophysica Acta* 990, 87-92.

Halliwell, B., Gutteridge, J.M.C., 1999. *Free Radicals in Biology and Medicine*. 3rd edition, Oxford University Press Inc., New York, NY, USA.

Haugland, R.P., 2005. *The Handbook. A Guide to Fluorescent Probes and Labeling Technologies*. 10th edition, Invitrogen Corp., USA.

Imlay, J.A., 2003. Pathways of oxidative damage. *Annual Review in Microbiology* 57, 395-418.

Ishikawa, T., Takeda, T., Shigeoka, S., Hirayama, O., Mitsunaga, T., 1993. Hydrogen peroxide generation in organelles of Euglena gracilis. *Phytochemistry* 33, 1297-1299.

Knauer, K., Behra, R., Sigg, L., 1997. Effects of free Cu^{2+} and Zn^{2+} ions on growth and metal accumulation in freshwater algae. *Environmental Toxicology and Chemistry* 16, 220-229.

Knauer, K., Ahner, B., Xue, H. B., Sigg, L., 1998. Metal and phytochelatin content in phytoplankton from freshwater lakes with different metal concentrations. *Environmental Toxicology and Chemistry* 17, 2444-2452.

Küpper, H., Küpper, F., Spiller, M., 1996. Environmental relevance of heavy metal-substituted chlorophylls using the example of water plants. *Journal of Experimental Botany* 47, 259-266.

Leonard, S.S., Harris, G.K., Shi, X.L., 2004. Metal-induced oxidative stress and signal transduction. *Free Radical Biology and Medicine* 37, 1921-1942.

Li, M., Hu, C. W., Zhu, Q., Chen, L., Kong, Z.M., Liu, Z.L., 2006. Copper and zinc induction of lipid peroxidation and effects on antioxidant enzyme activities in the microalga *Pavlova viridis* (Prymnesiophyceae). *Chemosphere* 62, 565-572.

Lu, C.M., Zhang, J.H., 1999. Copper-induced inhibition of PSII photochemistry in cyanobacterium *Spirulina platensis* is stimulated by light. *Journal of Plant Physiology* 154, 173-178.

Macdowell, F.D., 1949. The effects of some inhibitors of photosynthesis upon the photochemical reduction of a dye by isolated chloroplasts. *Plant Physiology* 24, 462-480.

Makino, N., Sasaki, K., Hashida, K., Sakakura, Y., 2004. A metabolic model describing the H_2O_2 elimination by mammalian cells including H_2O_2 permeation through cytoplasmic and peroxisomal membranes: comparison with experimental data. *Biochimica et Biophysica Acta - General Subjects* 1673, 149-159.

Mallick, N., 2004. Copper-induced oxidative stress in the chlorophycean microalga *Chlorella vulgaris*: response of the antioxidant system. *Journal of Plant Physiology* 161, 591-597.

Manzl, C., Enrich, J., Ebner, H., Dallinger, R., Krumschnabel, G., 2004. Copper-induced formation of reactive oxygen species causes cell death and disruption of calcium homeostasis in trout hepatocytes. *Toxicology* 196, 57-64.

Mohanty, N., Vass, I., Demeter, S., 1989. Copper toxicity affects photosystem II electron transport at the secondary quinone acceptor, QB. *Plant Physiology* 90, 175-179.

Morel, N.M.L., Rueter, J.G., Morel, F.M.M, 1978. Copper toxicity to *Skeletonema costatum* (Bacillariophyceae). *Journal of Phycology* 14, 43-48.

Morelli, E., Scarano, G., 2004. Copper-induced changes of non-protein thiols and antioxidant enzymes in the marine microalga *Phaeodactylum tricornutum*. *Plant Science* 167, 289-296.

Nagalakshmi, N., Prasad, M.N.V., 1998. Copper-induced oxidative stress in Scenedesmus bijugatus: Protective role of free radical scavengers. *Bulletin of Environmental Contamination and Toxicology* 61, 623-628.

Okamoto, O.K., Colepicolo, P., 1998. Response of superoxide dismutase to pollutant metal stress in the marine dinoflagellate *Gonyaulax polyedra*. *Comparative Biochemistry and Physiology. Part C Toxicology & Pharmcology* 119, 67-73.

Pätsikkä, E., Aro, E.M., Tyystjärvi, E., 2001. Mechanism of copper-enhanced photoinhibition in thylakoid membranes. *Physiologia Plantarum* 113, 142-150.

Pinto, E., Sigaud-Kutner, T.C.S., Leitao, M.A.S., Okamoto, O.K., Morse, D., Colepicolo, P., 2003. Heavy metal-induced oxidative stress in algae. *Journal of Phycology* 39, 1008-1018.

Rama Devi, S., Prasad, M.N.V., 1998. Copper toxicity in *Ceratophyllum demersum* L. (Coontail), a free floating macrophyte: Response of antioxidant enzymes and antioxidants. *Plant Science* 138, 157-165.

Ratkevicius, N., Correa, J.A., Moenne, A., 2003. Copper accumulation, synthesis of ascorbate and activation of ascorbate peroxidase in *Enteromorpha compressa* (L.) Grev. (Chlorophyta) from heavy metal-enriched environments in northern Chile. *Plant Cell Environment* 26, 1599-1608.

Reth, M., 2002. Hydrogen peroxide as second messenger in lymphocyte activation. *Nature Immunology* 3, 1129-1134.

Rijstenbil, J.W., Derksen, J.W.M., Gerringa, L.J.A., Poortvliet, T.C.W., Sandee, A., van den Berg, M., van Drie, J., Wijnholds, J.A., 1994. Oxidative stress induced by copper: defense and damage in the marine planktonic diatom *Ditylum brightwellii*, grown in continuous cultures with high and low zinc levels. *Marine Biology* 119, 583-590.

Sandmann, G., Böger, P., 1980. Copper-mediated lipid peroxidation processes in photosynthetic membranes. *Plant Physiology* 66, 797-800.

Schreiber, U. 2001. *Dual-Channel Photosynthesis Yield Analyzer ToxY-PAM. Handbook of Operation.* 2nd edition, Heinz Walz GmbH, Effeltrich, Germany.

Schreiber, U., Müller, J.F., Haugg, A., Gademann, R., 2002. New type of dual-channel PAM chlorophyll fluorometer for highly sensitive water toxicity biotests. *Photosynthesis Research* 74, 317-330.

Schröder, W. P., Arellano, J. B., Bittner, T., Baron, M., Eckert, H. J., Renger, G., 1994. Flash-induced absorption spectroscopy studies of copper interaction with photosystem II in higher plants. *Journal of Biological Chemistry* 269, 32865-32870.

Schwarzenbach, R.P., Escher, B.I., Fenner, K., Hofstetter, T.B., Johnson, C.A., von Gunten, U., Wehrli, B., 2006. The challenge of micropollutants in aquatic systems. *Science* 313, 1072-1077.

Seaver, L. C., Imlay, J.A., 2001. Hydrogen peroxide fluxes and compartmentalization inside growing *Escherichia coli*. *Journal of Bacteriology* 183, 7182-7189.

Seth, R., Yang, S., Cho, S., Sabean, M., Roberts, E.A., 2004. In vitro assessment of copper-induced toxicity in the human hepatoma line, Hep G2. *Toxicology In Vitro* 18, 501-509.

Steemann Nielsen, E., Kamp-Nielsen, L., Wium-Andersen, S., 1969. The effect of deleterious concentrations of copper on the photosynthesis of *Chlorella pyrenoidosa*. *Physiologia Plantarum* 22, 1121-1133.

Stevens, S.E., Patterson, C.O.P., Myers, J., 1973. The production of hydrogen peroxide by blue-green algae: a survey. *Journal of Phycology* 9, 427-430.

Sunda, W.G. 1988. Trace metal interaction with marine phytoplankton. *Biological Oceanography* 6, 411-441.

Sunda, W.G., Guillard, R.R.L., 1976. Relationship between cupric ion activity and toxicity of copper to phytoplankton. *Journal of Marine Research* 34, 511-529.

Tripathi, B.N., Mehta, S.K., Amar, A., Gaur, J.P., 2006. Oxidative stress in *Scenedesmus* sp. during short- and long-term exposure to Cu^{2+} and Zn^{2+}. *Chemosphere* 62, 538-44.

Uribe, E.G., Stark, B., 1982. Inhibition of photosynthetic energy-conversion by cupric ion - Evidence for Cu^{2+}-coupling factor I interaction. *Plant Physiology* 69, 1040-1045.

Vermaas, W.F.J., Styring, S., Schröder, W.P., Andersson, B., 1993. Photosynthetic water oxidation: the protein framework. *Photosynthesis Research* 38, 249-263.

Weger, H.G., 1997. Interactions between Cu(II), Mn(II) and salicylhydroxamic acid in determination of algal peroxidase activity. *Phytochemistry* 46, 195-201.

Xue, H.B., Sigg, L., 1993. Free cupric ion concentration and Cu(II) speciation in a eutrophic lake. *Limnology and Oceanography* 38, 1200-1213.

Yruela, I., Montoya, G., Alonso, P.J., Picorel, R., 1991. Identification of the pheophytin-QA-Fe domain of the reducing side of the photosystem II as the Cu(II)-inhibitory binding site. *Journal of Biological Chemistry* 266, 22847-22850.

Yruela, I., Alfonso, M., Dezarate, I.O., Montoya, G., Picorel, R., 1993. Precise location of the Cu(II)-inhibitory binding site in higher plant and bacterial photosynthetic reaction centers as probed by light-induced absorption changes. *Journal of Biological Chemistry* 268, 1684-1689.

Yruela, I., Pueyo, J.J., Alonso, P.J., Picorel, R., 1996a. Photoinhibition of photosystem II from higher plants - Effect of copper inhibition. *Journal of Biological Chemistry* 271, 27408-27415.

Yruela, I., Gatzen, G., Picorel, R., Holzwarth, A.R., 1996b. Cu(II)-inhibitory effect on photosystem II from higher plants. A picosecond time-resolved fluorescence study. *Biochemistry* 35, 9469-9474.

Zepp, R.G., Skurlatov, Y.I., Pierce, J.T., 1987. *Algal-induced decay and formation of hydrogen peroxide in water: its possible role in oxidation of anilines by algae*, in: Photochemistry environmental aquatic systems, eds. Zika, R.G., Cooper, W.J., American Chemical Society, Washington, USA, pp. 215-25.

Chapter 3

Mixture toxicity of three photosystem II inhibitors (atrazine, isoproturon, and diuron) towards photosynthesis of freshwater phytoplankton studied in outdoor mesocosms

3.1 Abstract

Mixture toxicity of three herbicides with the same mode of action was studied in a long-term outdoor mesocosm study. Photosynthetic activity of phytoplankton as the direct target site of the herbicides was chosen as physiological response parameter. The three photosystem II (PSII) inhibitors atrazine, isoproturon, and diuron were applied as 30 % hazardous concentrations (HC_{30}), which we derived from species sensitivity distributions calculated on the basis of EC_{50} growth inhibition data. The respective herbicide mixture comprised 1/3 of the HC_{30} of each herbicide. Short-term laboratory experiments revealed that the HC_{30} values corresponded to EC_{40} values when regarding photosynthetic activity as the response parameter. In the outdoor mesocosm experiment, effects of atrazine, isoproturon, diuron and their mixture on the photosynthetic activity of phytoplankton were investigated during a five-week period with constant exposure and a subsequent five-month post-treatment period when the herbicides dissipated. The results demonstrated that mixture effects determined at the beginning of constant exposure can be described by concentration addition since the mixture elicited a comparable phytotoxic effect as the single herbicides. Declining effects on photosynthetic activity during the experiment might be explained by both a decrease in water herbicide concentrations and by the induction of community tolerance.

3.2 Introduction

Due to human activities, natural freshwater ecosystems are exposed to multi-component mixtures composed of potentially toxic and simultaneous acting substances. In contrast to most other organic chemicals, pesticides are introduced into the environment, mainly for crop protection, with the intention to exert toxic effects on target organisms. Surface water contamination with pesticides originates from different sources, e.g., spray drift, surface run-off, drainage or accidental spills. Chemical analyses of surface waters throughout the world have demonstrated the widespread occurrence of pesticide mixtures in various freshwater bodies (Kreuger et al. 1998, Müller et al.

2002, Irace-Guigand et al. 2004, Leu et al. 2004, Kotrikla et al. 2006). In surface waters, the three pesticides atrazine, isoproturon, and diuron are often detected (Nitschke and Schüssler 1998, Graymore et al. 2001, Field et al. 2003) and are occurring in mixtures especially in spring (Kreuger et al. 1998, Irace-Guigand et al. 2004). These selective herbicides are widely used to control annual grass and broadleaf weeds and exert the same mode of toxic action as they interfere with the electron transport in photosystem II (PSII) by competing with plastoquinon for binding to the D1 protein in the thylakoid membrane (Trebst 1987).

A core problem for the assessment of the joint toxicity of pesticide mixtures in the context of setting environmental quality standards is that environmental concentrations of pollutants greatly vary on the spatial as well as on the temporal scale. Experimental testing of all conceivable combinations of pesticides is thus not feasible. However, if the mixture composition is known, the joint toxicity can be predicted from the toxicities of the single constituents. Two basic models, concentration addition and response addition also referred to as independent action are generally applied in predictive hazard assessment (Warne 2003). When the respective chemicals have the same mode of action and do not interact, concentration addition is assumed. The combined effect of such a mixture remains constant when one constituent is partially or totally replaced by the equieffective amount of another.

Previous mixture toxicity studies focused their investigations on single species (Hermens et al. 1982, Deneer et al. 1988, Altenburger et al. 2000, Faust et al. 2001) and natural communities (Arrhenius et al. 2004, Backhaus et al. 2004a, Arrhenius et al. 2006) in laboratory experiments. To investigate the applicability of the mixture toxicity concepts for the environment, studies should be performed as long-term studies in freshwater model ecosystems, such as enclosures, micro- and mesocosms. The effects of mixtures composed of similar or different types of pesticides were investigated in several freshwater model ecosystems (e.g., Hoagland et al. 1993, Fairchild et al. 1994, Hartgers et al. 1998, Beauvais et al. 1999, Cuppen et al. 2002, Wendt-Rasch et al. 2003, van Wijngaarden et al. 2004). However, to our knowledge, none of these studies focused on the suitability of the mixture toxicity concepts under environmental realistic conditions. In fact, the main purpose of these studies was to assess the potential ecological impact of realistic exposure scenarios on and to evaluate standards to be protective for the aquatic environment.

We performed an outdoor mesocosm study in 2006 to evaluate if effects of a mixture of similar acting compounds on the photosynthetic activity of phytoplankton can be described by concentration addition. Equipotent concentrations of the three PSII inhibitors atrazine (triazine), isoproturon, and diuron (both phenylurea derivates) and an equipotent mixture were applied for a constant exposure period of five weeks. In a subsequent post-treatment period of five months, the dissipation of the herbicides was quantified and discussed in relation to the photosynthetic activity of the phytoplankton community.

3.3 Materials and methods

3.3.1 Test chemicals

Atrazine (2-chloro-4-ethylamino-6-isopropyl-amino-s-triazine, CAS number 1912-24-9) (99 % purity) and diuron (3-(3,4-dichlorophenyl)-1,1-dimethylurea, CAS number 330-54-1) (purity 98.4 %) were provided by Syngenta (Basel, Switzerland) and DuPont Crop Protection (Newark, DE, USA), respectively. Isoproturon [3-(4-isopropylphenyl)-1,1-dimethylurea] (CAS number 34123-59-6) (analytical standard) was purchased from Sigma-Aldrich (Buchs, Switzerland).

3.3.2 Determination of test concentration

As a first step in mixture toxicity studies, concentration effect relationships for each single mixture compound need to be known for equipotent concentration ratios and are thus a prerequisite to make predictions by concentration addition and response addition for combined effects of a mixture with known constituent concentrations. In a mesocosm experiment such preliminary investigations are not feasible since number of test concentrations and replicates are strongly limited. Therefore, we have chosen a simplified mixture design and investigated the effects of equitoxic concentrations of the single herbicides and an equitoxic mixture on photosynthesis of phytoplankton. In the mixture each herbicide was present at the same fraction of its own individual toxicity. To approximate equitoxic concentrations, the 30 % hazardous concentrations (HC_{30}) were selected which were derived from species sensitivity distribution (SSD) curves comprising the results of a number of laboratory experiments investigating growth of a broad range of phototrophic organisms. We expected that the individual HC_{30} values of the single herbicides might induce similar effects on photosynthesis of phytoplankton. If the concentration addition concept holds true, the herbicide mixture with the following composition

$$c_{mix} = \frac{1}{3}(HC_{30_{atrazine}} + HC_{30_{isoproturon}} + HC_{30_{diuron}})$$

would elicit the same toxic effect on photosynthetic activity of phytoplankton as the HC_{30} of one of the three herbicides alone. SSD curves were constructed by ranking the EC_{50} values of 30 (atrazine), 11 (isoproturon) and 7 (diuron) phototrophic species in ascending order and estimating the percentiles as $(100 \times \text{rank})/(n+1)$, where n is the number of species in the distribution. EC_{50} data were collected from a literature compilation (Chèvre et al. 2006). Logarithmic hazardous concentrations for 30 % of the species (HC_{30}) were calculated on the basis of the logistic distribution of the EC_{50} values using Microcal software Origin™ 5.0 (Aldenberg and Slob 1993). Data were compared to calculations for the three herbicides using the approach described by Chèvre et al. (2006). Based on these results, HC_{30} were defined to be 70 µg/L = 325 nM for atrazine, 14 µg/L = 66 nM for isoproturon, and 5 µg/L = 21 nM for diuron. From SSDs provided by Schmitt-Jansen and Altenburger (2005) comparable HC_{30} for atrazine and isoproturon could be derived. The HC_{30} has been chosen to observe statistically significant effects, as variability in field experiments was assumed to be higher compared to laboratory investigations. To elucidate if the chosen HC_{30} elicit similar effects on photosynthetic activity of the phytoplankton from the mesocosms, short-term dose-response-relationships were established in the laboratory. Therefore, untreated phytoplankton was collected from the mesocosms and 5-fold concentrated by centrifugation. Algal samples of 3 mL were then dosed with four different concentrations of the three herbicides, i.e. atrazine (12.5, 25, 50 and 100 µg/L), isoproturon (2.5, 5, 10 and 20 µg/L), and diuron (1.25, 2.5, 5 and 10 µg/L). These concentrations were chosen to encompass the range of herbicide concentrations applied in the field (i.e., 1/3 HC_{30} and HC_{30}). Control samples received the same amount of the solvent methanol (0.1 %) as the herbicide-exposed ones. Exposure lasted for 3 h under an illumination of 40 µmol m^{-2} s^{-1}, at 25 °C and 100 rpm. Effects on photosynthesis were determined as described below.

3.3.3 Mesocosm experiment

The mesocosm test site of Syngenta Crop Protection AG was located in CH-8260 Stein, Switzerland (47°33'10" N, 7°57'47" E, 300 m a.s.l.). The site contains twenty-eight tanks made of high-density polyethylene which were buried in the ground to minimize rapid temperature fluctuations. The surface of each tank had a diameter of 3 m, a soil-sediment layer and water column of approximately 15 and 130 cm, respectively, yielding a volume of about 10 m^3. Phytoplankton, zooplankton, macroinvertebrates, and other organisms were introduced into the mesocosms along with the water and sediment from a nearby man made supply pond (~ 500 m^3, 4-m depth) and via aerial colonization during the year. Mesocosms were dominated by *Elodea canadensis* (Hydrocharitaceae) populations which were planted in the sediment.

Circulation was started in March, when all ponds were thawed, lasting for approximately six weeks, and then stopped before exposure was started. Herbicide concentrations and effects on photosynthesis of phytoplankton were investigated over a period of seven months from April until October 2006, which was divided into a six-week pre-exposure, a five-week constant exposure, and a five-month post-treatment period. Prior to application, mesocosms were randomly assigned to the different treatments. Each treatment and the control were replicated in three ponds. Herbicides were dissolved in methanol whereby the final solvent concentration in the ponds after application was negligible (< 0.002 %). Exposure started on 3 May 2006 (day 0). After application (day 0), herbicide concentrations in the single and mixture treatments were kept constant over a period of five weeks (day 0-34), i.e., the herbicides were supplemented to the ponds if necessary to maintain target herbicide concentrations at ± 20 %. Application of the herbicides was performed as spray drift on the water surface. The water column was mixed well with a polyethylene tube to achieve a homogenous distribution of the test substances. Separated sampling equipment such as tubes, buckets, etc. was used for each individual treatment to avoid cross contamination.

Sampling for chemical water analysis to control herbicide concentrations started on the day of first application and was performed twice a week during the period of constant exposure. Concentrations to be reapplied were calculated based on these measurements. In the post-treatment period the dissipation of the herbicides in the water column was analytically pursued by weekly measurements from day 34 to 68 and biweekly measurements from day 68 to 173. Biological sampling of phytoplankton was done in weekly intervals from day -8 to 54 and then in biweekly intervals from day 54 to 173. For chemical analysis and biological sampling depth-integrated water samples were taken from the four quadrants of each mesocosm using a polyethylene tube, 120 cm long and 4.5 cm in diameter. The tube was lowered close to the sediment surface to avoid sample contamination due to perturbation of the sediment. These raw water samples were used for chemical water analysis. For biological analysis, phytoplankton was separated from the zooplankton with an Apstein plankton net (Hydrobios Apparatebau GmbH, Kiel-Holtenau, Germany) of 25 cm in diameter and 55 µm mesh size. To receive reasonable maximum fluorescence yield values (F_m') ranging between 2500 and 3400 (Schreiber 2001) during determination of photosynthetic activity, phytoplankton samples (< 55 µm) were 5-fold concentrated in the laboratory. Therefore, a 50 ml phytoplankton sample from each pond was centrifuged at 3000 rpm for 5 min at room temperature and 40 ml of the supernatant was discarded. The algal pellet was resuspended in the remaining volume of 10 ml pond water.

3.3.4 Analytical methods

Samples were stored at 4 °C until analysis with online-solid phase extraction-liquid chromatography-tandem mass spectrometry SPE-HPLC-MS-MS. In brief, 18 ml filtrated water sample (cellulose acetate, 0.45 µm) were adjusted to pH 4 with acetate buffer, and isotopic-labeled pesticides were added as internal standards. Enrichment and elution of pesticides on a strata X column (2.1 mm I.D. x 20 mm I, Phenomenex) was carried out with an online-system using methanol for elution (Stoob et al. 2005). Separation was achieved by an Xbridge C18 column (2 mm I.D. x 50 mm L, Waters) at room temperature with a flow rate of 200 µL/min. The mobile phase was water with 0.1 % formic acid and methanol with 0.1 % formic acid, using an gradient of 40 to 90 % methanol within 20 min. MS-MS detection (TSQ Quantum, Thermo Electrons, San Jose, CA, USA) was performed in the selected reaction mode (SRM) with positive electrospray ionisation using a source voltage of 4.5 kV, ion transfer capillary temperature of 350 °C, sheath gas flow of 0.6 L/min and auxillary gas flow of 1.5 L/min. The monitored transitions of the pesticides and their internal standards are for atrazine: m/z 216 → 174, 104; for [D_5]-atrazine: m/z 221 -> 179, 101; for diuron m/z 233 → 72, 46; for [D_6]-diuron: m/z 239→ 78, 52; for isoproturon: m/z 207 → 72, 46; and for [D_6]-isoproturon: m/z 213 → 78, 52). Limits of quantification using signal-to-noise ratio >10 were 3 ng/L. As shown by Stoob et al. (2005) the intraday precision for the pesticides used in this study is 1-3 % and the recovery of the enrichment step 92 - 115 %. This results in an overall analytical error below 10 %. Besides the parent compounds the first known transformation products desethylatrazine (monitored transition m/z 188 → 146, 104), and despropylatrazine (m/z 174 → 104, 68) from atrazine, 1-(3,4-dichlorophenyl)-3-methylurea (DCPMU, m/z 219 → 162, 127) and 3,4-dichlorphenyl-urea (DCPU, m/z 205 → 162, 127) from diuron, and 3-(4-isopropylphenyl)-1-methylurea (IPPMU, m/z 193 → 94, 77) from isoproturon were analyzed. Their measurement started seven weeks after first application and lasted until the end of the study. Quantification was performed using the internal standards of the parent compounds.

3.3.5 Effects on photosynthesis

Effects on photosynthetic activity of phytoplankton were determined with a Toxy-PAM fluorometer (Heinz Walz GmbH, Effeltrich, Germany) by using the saturating pulse method (Schreiber et al. 2002). One mL of the 5-fold concentrated algal suspension described above was transferred into the quartz glass cuvette of the Toxy PAM fluorometer and PSII quantum yield Y was determined according to the following equation:

$$Y = \frac{F_m' - F}{F_m'}$$

where F is the momentary fluorescence yield measured between the pulses and F_m' is the maximum fluorescence yield induced by a saturation pulse leading to full inhibition of energy conversion at PSII reaction centres. In each algal sample five readings were recorded in 30 s distances and the mean of the last three Y values was used for further calculations. To present effects on phytoplankton, arithmetic means of the PSII quantum yield values of the three replicated ponds were determined and inhibition of photosynthetic activity (I) was calculated:

$$I\,[\%] = \left(1 - \frac{Y_{application}}{Y_{control}}\right) 100\%$$

Experimental data from laboratory short-term tests regarding herbicide-induced inhibition of photosynthetic activity of phytoplankton were analyzed using a linear regression of the GraphPad Prism 4.0 software (GraphPad, San Diego, CA, USA).

$$I(c) = m \log(c) + n$$

where I denotes the inhibition of photosynthetic activity (%), c is the herbicide concentration, m is a slope parameter and n determines the y-axis-intercept.

For the period of constant exposure (day 0-34) of the mesocosm experiment, time-weighted average effects on photosynthesis of phytoplankton were calculated as mean ± standard deviation of all measuring data points for each treatment.

3.3.6 Calculation of toxic units and half-life periods

To compare effects on photosynthesis between the various treatments, relative herbicide concentrations in the single herbicide treatments were expressed as toxic units (TU_i):

$$TU_i = \frac{c_i}{HC_{30_i}}$$

where c_i was the measured concentration of the herbicide i in the water column and the HC_{30i} was the 30 % hazardous effect concentration of the respective herbicide i obtained from SSD curves. According to the CA concept, toxic units in the mixture treatment (TU_{mix}) were calculated as:

$$TU_{mix} = \sum_{i=1}^{n} \frac{1}{n} \frac{c_i}{HC_{30_i}}$$

where n was the number of herbicides in the mixture, c_i/n was the concentration of the individual herbicide i in the mixture treatment and the HC_{30i} was the 30 % hazardous effect concentration of the respective herbicide i obtained from SSD curves.

Time-weighted average herbicide concentrations during the period of constant exposure (day 0-34) were calculated as mean ± standard deviation of actually measured concentrations estimated from all measuring data during the period of constant exposure.

The dissipation of the herbicides in the post-treatment period was described by calculating the half-life period ($t_{1/2}$) for the respective herbicide on the base of analytical data from single and mixture treatments according to first order kinetics:

$$\ln(c) = -kt + \ln(c_0)$$

$$t_{1/2} = \frac{\ln(2)}{k}$$

where t is the time, c is the herbicide concentration, c_0 is the herbicide concentration at the time t = 0, and k is the first order reaction rate.

3.4 Results

3.4.1 Herbicide concentrations in the water column

Relative water herbicide concentrations of atrazine, isoproturon, and diuron expressed as TU in the three single herbicide and the mixture treatment determined during the entire experiment (day 0-173) are shown in Fig. 3.1. Herbicide concentrations in the triplicate mesocosms receiving the same treatment were similar at the same sampling dates except for isoproturon. The average coefficients of variation (arithmetic mean ± standard deviation) were 4.5 ± 3.0 % (atrazine, n = 20), 16.7 ± 15.7 % (isoproturon, n = 24), 5.0 ± 3.7 % (diuron, n = 21) and 4.4 ± 3.2 % (mixture, n = 24). Reapplication of herbicides was performed up to three times during constant exposure to maintain the target HC_{30}. As it is shown in Fig. 3.1, atrazine was only supplemented once on day 20. Isoproturon concentrations rapidly decreased and had thus to be redosed three times (days 12, 20, 29). The mesocosms treated with diuron were redosed twice on day 12 and 20. In the mixture, all three herbicides were individually supplemented on day 12, 20 and 29.

During the period of constant exposure (day 0-34) averaged herbicide concentrations were determined to be in the range of target concentrations ± 20 % in all mesocosms for all herbicides (Table 3.1). Aimed target concentrations ± 20 % were exceeded and undershoot only in the mixture and isoproturon treatment on day 0 and 12, respectively (Table 3.1, Fig. 3.1).

In the post-exposure period (day 34-173), the dissipation of the herbicides was described by first order kinetics (Fig. 3.1). Half-life of atrazine ($t_{1/2}$ = 107 d, r^2 = 0.92) was approximately 3 and 2 times longer in comparison to those of isoproturon ($t_{1/2}$ = 35 d, r^2 = 0.80) and diuron ($t_{1/2}$ = 43,

$r^2 = 0.97$), respectively. At the end of the experiment (day 173), atrazine concentrations in the single treatment were still high with 0.42 ± 0.02 TU (n = 3) while diuron (0.11 ± 0.01 TU, n = 3) and isoproturon (0.08 ± 0.03 TU, n = 3) had nearly completely disappeared.

During the experiment, concentrations of the atrazine metabolites desethylatrazine and desisopropylatrazine did not exceed 11 and 6 % of the target concentration of the parent compound, respectively. IPPMU, the major metabolite of isoproturon, reached 5 % of the parent target herbicide concentration. Maximum DCPMU concentrations were 15 % whereas the other diuron metabolite DCPU could not be detected.

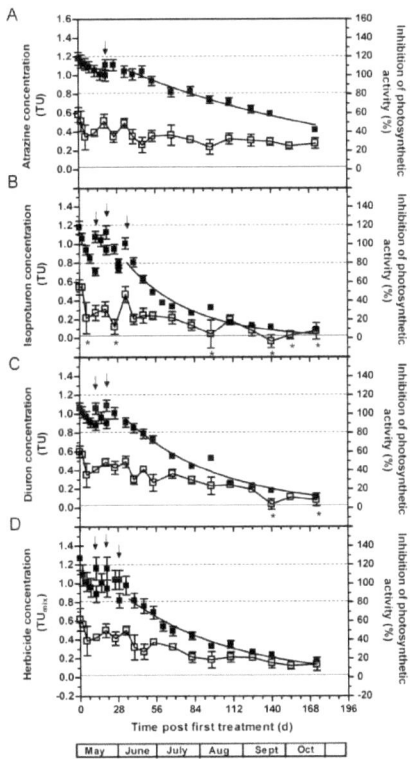

Fig. 3.1 Relative herbicide concentrations in the water phase (expressed as TU) (■) and inhibition of photosynthetic activity of phytoplankton (□) over time for the single treatments atrazine (A), isoproturon (B), diuron (C), and the mixture treatment (D). In general, data points represent arithmetic means ± standard deviations. Standard deviations were estimated according to Gaussian error propagation. Dashed lines indicate target concentrations ± 20 %. Arrows mark reapplications of the herbicides. Inhibition of photosynthetic activity of phytoplankton that was not significantly different from control level is indicated by an asterisk (two-tailed, unpaired t-test, $p > 0.05$).

Table 3.1 Minimum, maximum and average herbicide concentrations (expressed as TU) as well as averaged inhibition of photosynthetic activity of phytoplankton (in %) calculated for the period of constant exposure (day 0-34).

	Atrazine	Isoproturon	Diuron	Mixture
Minimum and maximum herbicide concentration (TU)	0.95 to 1.22	0.43 to 1.20	0.86 to 1.14	0.75 to 1.32
Average herbicide concentration (TU)	1.09 ± 0.07 (n = 30)	0.96 ± 0.17 (n = 39)	0.98 ± 0.08 (n = 33)	1.03 ± 0.13 (n = 39)
Inhibition of photosynthetic activity (%)	45.6 ± 9.3 (n = 7)	35.6 ± 16.7 (n = 7)	47.7 ± 9.0 (n = 7)	48.6 ± 8.2 (n = 7)

3.4.2 Effects on photosynthesis

Laboratory short-term toxicity tests

Herbicide-induced inhibition of photosynthesis of the phytoplankton sampled from untreated mesocosms is presented in Fig. 3.2. The shape and slope of the concentration-response-curves of the three herbicides in the selected concentration range were similar. The HC_{30} of atrazine, isoproturon, and diuron elicited similar effects accounting for 43.0 ± 7.6, 36.2 ± 6.9 and 40.6 ± 5.8 % inhibition of photosynthesis (n = 6), respectively (one way ANOVA, p > 0.05). Effects induced by 1/3 HC_{30} of atrazine, isoproturon, and diuron corresponded to 22.7 ± 3.3 (n = 5), 16.1 ± 3.2 (n = 6), and 17.5 ± 6.6 % (n = 6) inhibition of photosynthesis and were comparable (one way ANOVA, p > 0.05). The effects induced by the $HC_{30\ Isoproturon}$ and by 1/3 $HC_{30\ Isoproturon}$ were statistically significant different from each other (two-tailed unpaired t-test, p < 0.05). This held also true for the other two herbicides.

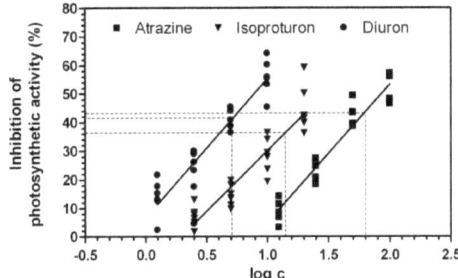

Fig. 3.2 Inhibition of photosynthetic activity (%) of phytoplankton due to the three single herbicides atrazine (■), isoproturon (▼), and diuron (●) determined in short-term laboratory experiments. Experimental data (■, ▼, ●) and statistical fit (—) according to a linear regression are shown. Logarithmic herbicide concentrations were calculated from concentration values expressed in µg/L. Dashed lines indicate the effect levels that correspond to the HC_{30} of the three herbicides.

Mesocosm experiment

Prior to application (day -8) photosynthetic activity of phytoplankton in all mesocosm was comparable (Y = 0.457 ± 0.008, n = 15). During spring (day 0-68), yield values increased to 0.602 ± 0.020 (n = 3) in control ponds due to a higher photosynthetic activity of the algal community and were constant during summer (day 68-173) (Y = 0.571 ± 0.034, n = 24). Within two days after application (day 2) exposure to atrazine, isoproturon, diuron, and the mixture induced a similar inhibition of photosynthetic activity of phytoplankton accounting for 51.4 ± 10.8 %, 54.8 ± 3.5 %, 57.1 ± 3.3 %, and 56.3 ± 3.0 % (n = 3 mesocosms), respectively (one way ANOVA, p > 0.05) (Fig. 3.3).

Inhibition of photosynthetic activity of phytoplankton in the single herbicide and mixture treatments, averaged over the period of constant exposure (day 0-34), was not significantly different from each other (one way ANOVA, p > 0.05) (Table 3.1). Mean effects induced by the HC_{30} of the single herbicides on photosynthesis during constant exposure of the mesocosm experiment and effects of the HC_{30} of atrazine, isoproturon, and diuron obtained from short term laboratory tests were similar as well (one way ANOVA, p > 0.05) although phytoplankton exposed in the field tended to be more strongly affected probably due to longer exposure time. During the entire mesocosm experiment (day 0-173), a linear trend between declining effects on photosynthesis and decreasing herbicide concentrations was observed (Fig. 3.4). Compared to control ponds, statistically significant effects on photosynthesis induced by atrazine could be determined until the end of the experiment (day 173), whereas effects of isoproturon and diuron completely disappeared

Chapter 3 – Mixture toxicity to photosynthesis of phytoplankton

on day 96 and 140, respectively (two-tailed unpaired t-test, p>0.05) (Fig. 3.1). In the mixture treatment, photosynthetic activity was still statistically significant reduced compared to control mesocosms at the end of the mesocosm experiment (Fig. 3.1). Overall, the effect parameter photosynthetic activity of phytoplankton measured as PSII quantum yield showed a low variability since measured data in the three replicated ponds were similar at the different sampling dates (coefficient of variation: 7.2 ± 4.9 %, n = 95).

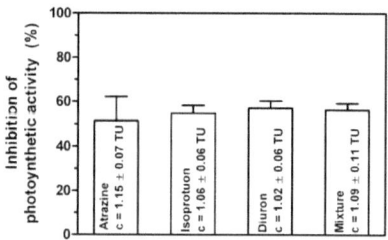

Fig. 3.3 Herbicide-induced inhibition of photosynthetic activity (%) determined on day 2 after application. Each bar represents mean and standard deviation of the three replicated mesocosms. Statistical analysis of the data (one way ANOVA; p > 0.05) revealed no significant differences among the single bars. Corresponding water herbicide concentrations (in TU) detected on day 2 were added.

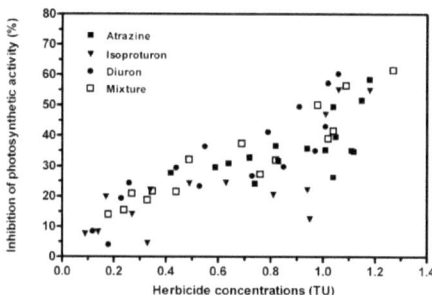

Fig. 3.4 Concentration-response relationship between inhibition of photosynthetic activity of phytoplankton (in %) and herbicide concentrations (expressed as TU) for atrazine (■), isoproturon (▼), diuron (●), and the mixture (□). Measured data from the entire mesocosm experiment are presented.

3.5   Discussion

For the first time mixture toxicity of three similar acting PSII inhibitors was studied under environmental conditions in a mesocosm study. Photosynthetic activity of phytoplankton was chosen as the response parameter since the mode of action of the selected PSII inhibitors implies a direct interference of the herbicide molecule with the photosynthetic apparatus.

Atrazine, isoproturon, and diuron concentrations equivalent to 30 % hazardous concentrations derived from growth based SSDs were demonstrated to be equipotent toxic when photosynthesis of phytoplankton was regarded as response parameter. This was true under laboratory as well as under field conditions. Moreover, effects of the herbicide mixture observed at the beginning of the mesocosm experiment can be described by concentration addition due to the fact that the mixture comprising 1/3 of the HC_{30} of each herbicide induced the same effect as the HC_{30} of the single herbicides. Thereby, we assume that the effects observed after two days of exposure were a direct result of the exposure concentrations and that changes in phytoplankton community structure as a consequence of herbicide-induced tolerance did not occur after 2 days of exposure. Our findings are consistent with results from laboratory studies demonstrating that mixtures of urea/triazine type PSII inhibitors follow concentration addition (Faust et al. 2001, Drost et al. 2003, Arrhenius et al. 2004, Backhaus et al. 2004b). These mixture toxicity studies investigated growth of single phototrophic species (Faust et al. 2001, Drost et al. 2003, Backhaus et al. 2004b) and inhibition of photosynthesis of marine periphyton and epipsammon communities (Arrhenius et al. 2004).

Inhibition of photosynthesis measured as *in vivo* chlorophyll fluorescence proved to be a reliable endpoint to evaluate mixture toxicity to phytoplankton under environmental conditions since variability determined for single exposure days, e.g., as shown for day 2, was low. Variability of the mean effect parameter examined over the entire period of constant exposure was higher at least in the case of isoproturon. This enhanced variability of the effect measurements was related to the variability of the mean exposure concentrations calculated from 11 sampling occasions during constant exposure and closely linked to the dissipation of the herbicides. For instance, isoproturon possessed the lowest half-life resulting in the highest variation of mean exposure concentrations and corresponding effects. In contrast, higher half-life estimates of atrazine and diuron were reflected in moderately variable mean concentrations and effects on photosynthesis.

In the post-treatment period, herbicide concentrations in the water column of the mesocosms dissipated according to first order kinetics. The half-lives of atrazine, isoproturon, and diuron corresponded well to half-life estimates from different other freshwater microcosm studies whereby slight discrepancies may be explained by differences in temperature, light source, or sediment type (Kemp et al. 1985, Rönnefahrt et al. 1997, Hartgers et al. 1998, Merlin et al. 2002). Dissipation of the two phenyl urea herbicides was mainly due to microbial degradation processes (Sørensen et al. 2001, Field et al. 2003). The slow dissipation of atrazine can be explained by a marginal degradation of this herbicide via biotic and abiotic pathways.

Effects on photosynthesis of the phytoplankton were demonstrated to be related to water herbicide concentrations. However, other factors such as herbicide-induced community tolerance might have also influenced the strength of the effects on photosynthesis and their reversibility. Differences in the sensitivities of the various algae towards the herbicides might have led to changes in community composition. As it has been discussed by Blanck (2002), chemical stressors exert a selection pressure that hinders the success of sensitive species whereby tolerant species become more dominant. Possible tolerance mechanism might be elevated numbers of PSII, an enhanced antioxidative response (Geoffroy et al. 2002), exclusion of the herbicide from the target site, metabolization of the herbicide (Tang et al. 1998), or the use of alternative energy sources (Heifetz et al. 2000). In the case of atrazine, community tolerance might have played a minor role only due to the fact that at the end of the experiment 0.4 TUs still resulted in a 25 % inhibition of photosynthesis. Isoproturon and diuron exposure did not result in sustained effects on photosynthesis of the algal communities since photosynthetic activity of exposed phytoplankton was comparable to that of untreated phytoplankton from day 140 on.

The metabolites of the herbicides did not substantially contribute to the total toxicity. This conclusion is supported by low metabolite concentrations (< 15 %) and toxicities except for DCPMU (US EPA ecotoxicology database, Escher et al. 2007). From laboratory experiments with *Pseudokirchneriella subcapitata* it can be concluded that this diuron metabolite possesses 75 % of the toxic potential of the parent (US EPA ecotoxicology database, Escher et al. 2007). However, for an algal community no data on the toxicity of DCPMU has been reported. Thus, we cannot exclude that DCPMU was contributing to the overall toxicity in the diuron and mixture treatments.

Overall, we demonstrated that concentration addition of individual mixture components can describe mixture effects observed under field conditions by focussing our investigations on photosynthesis of an algal community as a response parameter that corresponds to the direct target of the investigated substances. If the concept of concentration addition is also suitable to predict effects on response parameters that describe higher levels of biological organization, i.e. composition, structure, succession, or ecological functioning of the algal community, remains an interesting question that we are addressing in our ongoing research.

3.6 References

Aldenberg, T., Slob, W., 1993. Confidence-limits for hazardous concentrations based on logistically distributed NOEC toxicity data. *Ecotoxicology and Environmental Safety* 25, 48-63.

Altenburger, R., Backhaus, T., Boedeker, W., Faust, M., Scholze, M., Grimme, L.H., 2000. Predictability of the toxicity of multiple chemical mixtures to *Vibrio fischeri*: Mixtures composed of similarly acting chemicals. *Environmental Toxicology and Chemistry* 19, 2341-2347.

Arrhenius, A., Gronvall, F., Scholze, M., Backhaus, T., Blanck, H., 2004. Predictability of the mixture toxicity of 12 similarly acting congeneric inhibitors of photosystem II in marine periphyton and epipsammon communities. *Aquatic Toxicology* 68, 351-367.

Arrhenius, A., Backhaus, T., Gronvall, F., Junghans, M., Scholze, M., Blanck, H., 2006. Effects of three antifouling agents on algal communities and algal reproduction: Mixture toxicity studies with TBT, Irgarol, and Sea-Nine. *Archives of Environmental Contamination and Toxicology* 50, 335-345.

Backhaus, T., Arrhenius, A., Blanck, H., 2004a. Toxicity of a mixture of dissimilarly acting substances to natural algal communities: Predictive power and limitations of independent action and concentration addition. *Environmental Science & Technology* 38, 6363-6370.

Backhaus, T., Faust, M., Scholze, M., Gramatica, P., Vighi, M., Grimme, L. H., 2004b. Joint algal toxicity of phenylurea herbicides is equally predictable by concentration addition and independent action. *Environmental Toxicology and Chemistry* 23, 258-264.

Beauvais, S. L., Atchison, G. J., Stenback, J.Z., Crumpton, W.G., 1999. Use of cholinesterase activity to monitor exposure of *Chironomus riparius* (Diptera : Chironomidae) to a pesticide mixture in hypoxic wetland mesocosms. *Hydrobiologia* 416, 163-170.

Blanck, H., 2002. A critical review of procedures and approaches used for assessing pollution-induced community tolerance (PICT) in biotic communities. *Human and Ecological Risk Assessment* 8, 1003-1034.

Chèvre, N., Loeppe, C., Singer, H., Stamm, C., Fenner, K., Escher, B.I., 2006. Including mixtures in the determination of water quality criteria for herbicides in surface water. *Environmental Science & Technology* 40, 426-435.

Cuppen, J.G.M., Crum, S.J.H., Van den Heuvel, H.H., Smidt, R.A., Van den Brink, P.J., 2002. Effects of a mixture of two insecticides in freshwater microcosms: I. Fate of chlorpyrifos and lindane and responses of macroinvertebrates. *Ecotoxicology* 11, 165-180.

Deneer, J.W., Seinen, W., Hermens, J.L.M., 1988. Growth of *Daphnia magna* exposed to mixtures of chemicals with diverse modes of action. *Ecotoxicology and Environmental Safety* 15, (1), 72-77.

Drost, W., Backhaus, T., Vassilakaki, M., Grimme, L.H., 2003. Mixture toxicity of s-triazines to Lemna minor under conditions of simultaneous and sequential exposure. *Fresenius' Environmental Bulletin.* 12, 601-607.

Escher, B.I., Baumgartner, R., Lienerts, J., Fenner, K., 2007. *Predicting the Ecotoxicological Effects of Transformation Products,* in: Handbook of Environmental Chemistry – Degradation of Synthetic Chemicals in the Environment, eds. Boxall, A., Springer Verlag, Heidelberg, Germany, in press.

Fairchild, J.F., Lapoint, T.W., Schwartz, T.R., 1994. Effects of an herbicide and insecticide mixture in aquatic mesocosms. *Archives of Environmental Contamination and Toxicology* 27, 527-533.

Faust, M., Altenburger, R., Backhaus, T., Blanck, H., Boedeker, W., Gramatica, P., Hamer, V., Scholze, M., Vighi, M., Grimme, L.H., 2001. Predicting the joint algal toxicity of multi-component s-triazine mixtures at low-effect concentrations of individual toxicants. *Aquatic Toxicology* 56, 13-32.

Field, J.A., Reed, R.L., Sawyer, T.E., Griffith, S.M., Wigington, P.J., 2003. Diuron occurrence and distribution in soil and surface and ground water associated with grass seed production. *Journal of Environmental Quality* 32, 171-179.

Geoffroy, L., Teisseire, H., Couderchet, M., Vernet, G., 2002. Effect of oxyfluorfen and diuron alone and in mixture on antioxidative enzymes of *Scenedesmus obliquus*. *Pesticide Biochemistry and Physiology* 72, 178-185.

Graymore, M., Stagnitti, F., Allinson, G., 2001. Impacts of atrazine in aquatic ecosystems. *Environmental International* 26, 483-495.

Hartgers, E.M., Aalderink, G.H.R., Van den Brink, P.J., Gylstra, R., Wiegman, J.W.F., Brock, T.C.M., 1998 Ecotoxicological treshold levels of a mixture of herbicides (atrazine, diuron and metolachlor) in freshwater microcosms. *Aquatic Ecology* 32, 135-152.

Heifetz, P.B., Forster, B., Osmond, C.B., Giles, L.J., Boynton, J.E., 2000. Effects of acetate on facultative autotrophy in *Chlamydomonas reinhardtii* assessed by photosynthetic measurements and stable isotope analyses. *Plant Physiology* 122, 1439-1445.

Hermens, J., Leeuwangh, P., 1982. Joint toxicity of mixtures of 8 and 24 chemicals to the guppy (*Poecilia reticulata*). *Ecotoxicology and Environmental Safety* 6, (3), 302-310.

Hoagland, K.D., Drenner, R.W., Smith, J.D., Cross, D.R., 1993. Fresh-water community responses to mixtures of agricultural pesticides - Effects of atrazine and bifenthrin. *Environmental Toxicology and Chemistry* 12, 627-637.

Irace-Guigand, S., Aaron, J.J., Scribe, P., Barcelo, D., 2004. A comparison of the environmental impact of pesticide multiresidues and their occurrence in river waters surveyed by liquid chromatography coupled in tandem with UV diode array detection and mass spectrometry. *Chemosphere* 55, 973-981.

Kemp, W.M., Boynton, W.R., Cunningham, J.J., Stevenson, J.C., Jones, T.W., Means, J.C., 1985. Effects of atrazine and linuron on photosynthesis and growth of the macrophytes, *Potamogeton perfoliatus* L. and *Myriophyllum-Spicatum* L. in an estuarine environment. *Marine Environmental Research* 16, 255-280.

Knauer, K., Maise, S., Thoma, G., Hommen, U., Gonzalez-Valero, J., 2005. Long-term variability of zooplankton populations in aquatic mesocosms. *Environmental Toxicology and Chemistry* 24, 1182-1189.

Kotrikla, A., Gatidou, G., Lekkas, T. D., 2006. Monitoring of triazine and phenylurea herbicides in the surface waters of Greece. *Journal of Environmental Sciences and Health Part B Pesticides Food Contaminants and Agricultural Wastes* 41, 135-144.

Kreuger, J., 1998. Pesticides in stream water within an agricultural catchment in southern Sweden, 1990-1996. *Science of the Total Environment* 216, 227-251.

Leu, C., Singer, H., Stamm, C., Müller, S.R., Schwarzenbach, R.P., 2004. Simultaneous assessment of sources, processes, and factors influencing herbicide losses to surface waters in a small agricultural catchment. *Environmental Science & Technology* 38, 3827-3834.

Merlin, G., Vuillod, M., Lissolo, T., Clement, B., 2002. Fate and bioaccumulation of isoproturon in outdoor aquatic microcosms. *Environmental Toxicology and Chemistry* 21, 1236-1242.

Müller, K., Bach, M., Hartmann, H., Spiteller, M., Frede, H.G., 2002. Point- and nonpoint-source pesticide contamination in the Zwester Ohm catchment, Germany. *Journal of Environmental Quality* 31, 309-318.

Nitschke, L., Schüssler, W., 1998. Surface water pollution by herbicides from effluents of waste water treatment plants. *Chemosphere* 36, 35-41.

Rönnefahrt, I., Traub-Eberhard, U., Kordel, W., Stein, B., 1997. Comparison of the fate of isoproturon in small- and large-scale water/sediment systems. *Chemosphere* 35, 181-189.

Schmitt-Jansen, M., Altenburger, R., 2005. Predicting and observing responses of algal communities to photosystem II-herbicide exposure using pollution-induced community tolerance and species-sensitivity distributions. *Environmental Toxicology and Chemistry* 24, 304-312.

Schreiber, U., 2001. *Dual-Channel Photosynthesis Yield Analyzer ToxY-PAM. Handbook of Operation.* 2nd edition. Heinz Walz GmbH, Effeltrich, Germany.

Schreiber, U., Müller, J.F., Haugg, A., Gademann, R., 2002. New type of dual-channel PAM chlorophyll fluorometer for highly sensitive water toxicity biotests. *Photosynthesis Research* 74, 317-330.

Sørensen, S.R., Ronen, Z., Aamand, J., 2001. Isolation from agricultural soil and characterization of a *Sphingomonas* sp able to mineralize the phenylurea herbicide isoproturon. *Applied and Environmental Microbiology* 67, 5403-5409.

Stoob, K., Singer, H.P., Goetz, C.W., Ruff, M., Müller, S.R., 2005. Fully automated online solid phase extraction coupled directly to liquid chromatography-tandem mass spectrometry - Quantification of sulfonamide antibiotics, neutral and acidic pesticides at low concentrations in surface waters. *Journal of Chromatography A* 1097, 138-147.

Tang, J.X., Siegfried, B.D., Hoagland, K.D., 1998. Glutathione-S-transferase and in vitro metabolism of atrazine in freshwater algae. *Pesticide Biochemistry and Physiology* 59, 155-161.

Trebst, A., 1987. The 3-dimensional structure of the herbicide binding niche on the reaction center polypeptides of photosystem II. *Zeitschrift für Naturforschung - Journal of Biosciences* 42, 742-750.

U.S. EPA ecotoxicology database, http://www.epa.gov/ecotox/

van Wijngaarden, R.P.A., Cuppen, J.G.M., Arts, G.H.P., Crum, S.J.H., van den Hoorn, M.W., van den Brink, P.J., Brock, T.C.M., 2004. Aquatic risk assessment of a realistic exposure to pesticides used in bulb crops: A microcosm study. *Environmental Toxicology and Chemistry* 23, 1479-1498.

Warne. M.S.J., 2002. *A review of the ecotoxicity of mixtures, approaches to, and recommendations for, their management,* in: *Proceedings of the Fifth National Workshop on the Assessment of*

Site Contamination, eds. Langley, A., Gilbey, M., Kennedy, B., NEPC Service Corporation, Adelaide, Australia, pp 253-276.

Wendt-Rasch, L., Pirzadeh, P., Woin, P., 2003. Effects of metsulfuron methyl and cypermethrin exposure on freshwater model ecosystems. *Aquatic Toxicology* 63, 243-256.

Chapter 4

Effects of PSII inhibitors and their mixture on freshwater phytoplankton succession in outdoor mesocosms

4.1 Abstract

Effects of three photosystem II inhibitors and of their mixture on a freshwater phytoplankton community were studied in outdoor mesocosms. Atrazine, isoproturon, and diuron were applied as 30 % hazardous concentrations (HC_{30}) obtained from species sensitivity distributions. Taking the concept of concentration addition into account, the mixture comprised 1/3 of the HC_{30} of each substance. Effects were investigated during a five-week period of constant concentrations and a subsequent five-month post-treatment period when the herbicides dissipated. Community parameters such as total abundance, species composition, and diversity as well as recovery of the community were evaluated. Ordination techniques, such as the principal component analysis (PCA), were applied to compare the various treatments on community level. The three single herbicides stimulated comparable effects on total abundance and diversity of phytoplankton during the period of constant exposure due to the susceptibility of the dominant Cryptophytes *Chroomonas acuta* and *Cryptomonas erosa et ovata* and the Prasinophyte *Nephroselmis* cf. *olivacea*. Moreover, concentration addition was found to describe combined effects of atrazine, isoproturon, and diuron on total abundance and diversity in the constant exposure period since their mixture induced effects on abundance and diversity similar to the single substances. PCA revealed that community structure of diuron and isoproturon treated phytoplankton recovered two weeks after constant exposure which might be related to the fast dissipation of these phenylureas. The species composition of mixture and atrazine treated communities were not comparable to that of the control community five months after the end of the constant exposure period. This might be explained by the slow dissipation of atrazine and by differences in the species sensitivities resulting in a different succession of the phytoplankton.

4.2 Introduction

Chemical analysis of surface waters conducted throughout the world indicates that pollution of our aquatic ecosystems by pesticide contamination is a major environmental concern (Schwarzenbach et al. 2006). Pesticides mostly enter the aquatic environment via spray drift, drainage, and run-off events from agricultural areas, effluents from waste water treatment plants, or accidental spills. At the end of last century, many European and North American countries have defined specific quality criteria derived from predicted no effect concentrations for single pesticides in surface waters to protect the aquatic environment against pesticide impact (Stephan et al. 1985, Nowell et al. 1994, Zabel et al. 1999, Jahnel et al. 2001, Babut et al. 2003). With the EU Water Framework Directive (WFD, Directive 2000/60/EC) a strategy was adopted to harmonize the derivation of quality standards for 33 priority pollutants within the countries of the European Community. However, it has to be considered that mixtures instead of the presence of single pesticides may be the most common exposure scenario in the field. Estimation and regulation of the joint toxicity of pesticide combinations is therefore still a matter of debate. To address the problem of mixtures as the most prominent exposure scenario in surface waters, Chèvre et al. (2006) previously suggested a new approach to calculate surface water quality criteria for herbicide mixtures from ecotoxicological values derived from species sensitivity distributions by considering concentration addition. Concentration addition and independent action are two different concepts generally used to predict the joint toxicity of chemicals that do not affect the toxicity of one another: Concentration addition concerns mixtures of substances with the same mode of action whereas independent action is used to predict combined effects of toxicants with a dissimilar mode of action (Warne 2003). Effects of mixtures of herbicides on photosynthesis or growth of single algal species and algal communities were studied in several laboratory experiments (e.g., Arrhenius et al. 2004, 2006, Backhaus et al. 2004a, b, Drost et al. 2003, Faust et al. 2001, Junghans et al. 2003). Adverse effects on higher levels of biological organization i.e. on species composition and succession pattern of periphyton or phytoplankton communities and the potential of such algal communities to recover from chemical stress have been extensively studied, but usually for single herbicides only (e.g., Bérard et al. 2003, Dorigo et al. 2004, Gustavson et al. 2003, Peres et al. 1996). One exception is the work of Hartgers et al. (1998) who investigated combined effects of a mixture of three herbicides with two different modes of action on structural and functional endpoints in freshwater mesocosms. However, the aim of their study was not to assess the applicability of mixture toxicity concepts for the field situation but to evaluate if common procedures to calculate water quality standards for single herbicides from standard single algal test species such as *Pseudokirchneriella subcapitata* are sufficient to protect aquatic ecosystems against herbicide mixture effects.

Atrazine, isoproturon, and diuron are three herbicides and widely used to control annual grass and broadleaf weeds. They have often been detected simultaneously in surface waters especially in spring (Graymore et al. 2001, Field et al. 2003, Nitschke et al. 1998) and many studies have reported on their adverse effects on non-target organisms (Dorigo et al. 2004, Kirby et al. 1994, Solomon et al. 1996, Teisseire et al. 1999, Traunspurger et al. 1996). The phytotoxic mode of action of these three herbicides is based on the interference with the electron transport chain of photosystem II (PSII) of algae and higher plants. They bind specifically to the quinon site at the D1 protein within the thylakoid membrane resulting in a subsequent oxidative damage to the photosynthetic apparatus (Trebst 1987).

Previously, we demonstrated that concentration addition of atrazine, isoproturon, and diuron can describe the effects of their mixture on photosynthesis of phytoplankton communities under laboratory as well as under field conditions (Knauert et al. 2008). In the present study, we addressed the question whether equitoxic concentrations of the three herbicides and their equitoxic mixture result in similar responses of the phytoplankton community structure. Thus, herbicide effects were evaluated using principal component analysis (PCA) besides common community parameters such as total abundance, number of species, diversity, dominance patterns (with respect to the level of classes), and dynamics of the most dominant species.

4.3 Material and methods

4.3.1 Test chemicals

Atrazine (2-chloro-4-ethylamino-6-isopropyl-amino-s-triazine, CAS number 1912-24-9) (99 % purity) and diuron (3-(3,4-dichlorophenyl)-1,1-dimethylurea, CAS number 330-54-1) (purity 98.4 %) were supplied by Syngenta (Basel, Switzerland) and DuPont Crop Protection (Newark, DE, USA), respectively. Isoproturon [3-(4-isopropylphenyl)-1,1-dimethylurea] (CAS number 34123-59-6) (analytical standard) was purchased from Sigma-Aldrich (Buchs, Switzerland). Herbicides were dissolved in methanol (p.a., Merck, Darmstadt, Germany) prior to application whereby the final solvent concentration in the ponds was less than < 0.002%.

For testing the single herbicides, the three PSII inhibitors were applied as 30 % hazardous concentrations (HC_{30}) derived from growth based (EC_{50}) species sensitivity distributions (SSD) as described in detail in Knauert et al. (2008). Since the concept of concentration addition can be applied for these herbicides, the mixture was prepared from 1/3 of the HC_{30} of each of the three herbicides. Herbicide concentrations in the single substance treatments were 70 µg/L = 325 nM for atrazine, 14 µg/L = 66 nM for isoproturon, and 5 µg/L = 21 nM for diuron while in the mixture atrazine, isoproturon, and diuron accounted for 23.3 µg/L = 108 nM, 4.7 µg/L = 22 nM and 1.7 µg/L = 7 nM.

4.3.2 Mesocosm test site

The mesocosm test facility of Syngenta Crop Protection AG is located in CH-8260 Stein, Switzerland (47°33'10" N, 7°57'47" E, 300 m a.s.l.). The site of Syngenta contains twenty-eight tanks made of high-density polyethylene which were buried in the ground to minimize rapid temperature fluctuations. The surface of each tank had a diameter of 3 m, a soil-sediment layer and water column of approximately 15 and 130 cm, respectively, yielding a volume of about 10 m^3. Phytoplankton, zooplankton, macroinvertebrates, and other organisms were introduced into the mesocosms along with the water and sediment from a nearby man made supply pond (~ 500 m^3, 4-m depth) and via aerial colonization during the year. Mesocosms were dominated by *Elodea canadensis* (Hydrocharitaceae) populations which were planted in the sediment.

4.3.3 Application and exposure regime

During March to April 2006, water was circulated by pumping water between the mesocosms and the supply pond. Prior to application, water circulation was stopped and mesocosms were regarded as isolated systems. Mesocosms were randomly assigned to the different treatments. Each treatment and the control were replicated in three ponds. Exposure started on 3 May 2006 (day 0). Over a period of five weeks (day 0-34), the herbicide concentrations in the single and mixture treatments were kept constant by adding amounts of the herbicides to the ponds to maintain target herbicide concentrations at ± 20 %. The herbicides were applied to the water surface of the mesocosm by using a commercial sprayer. After application, the water column was carefully mixed with a polyethylene tube to achieve a homogenous distribution of the test chemicals.

The phytoplankton community as well as herbicide concentrations were monitored over a period of seven months from April to October 2006. The entire mesocosm experiment was thus subdivided into a six-week pre-exposure, a five-week constant exposure and a five-month post-treatment period (Fig. 4.1). Separated sampling equipment such as tubes, buckets, etc. was used for each individual treatment to avoid cross contamination.

4.3.4 Chemical and biological sampling

Sampling for chemical water analysis to determine herbicide concentrations started on the day of application. In the post-treatment period, the dissipation of the herbicides in the water column was analytically pursued. Biological sampling of phytoplankton was done in weekly intervals from day -8 to 54 and then in biweekly intervals from day 54 to 173. For chemical analysis and biological sampling, depth-integrated water samples were taken from the four quadrants of each mesocosm using a polyethylene tube, 120 cm long and 4.5 cm in diameter. The tube was lowered close to the bottom of the pond to avoid sample contamination due to perturbation of the sediment. These raw water samples were used for chemical water analysis. For taxonomic analysis, phytoplankton was separated from the zooplankton with an Apstein plankton net (Hydrobios Apparatebau GmbH, Kiel-Holtenau, Germany) of 25 cm in diameter and 55 µm mesh size. At the various sampling dates, 100 mL phytoplankton samples were taken from each mesocosm, stained and conserved with approximately 2 mL Lugol-iodine solution (Schwoerbel 2005) and stored in the dark for further taxonomic determination.

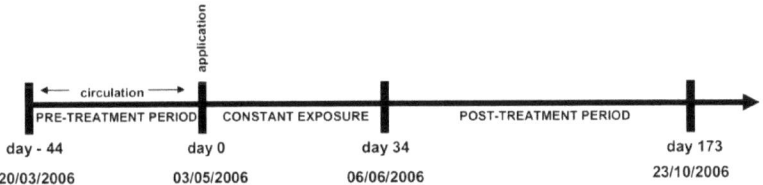

Fig. 4.1 Temporal scale of the mesocosm study.

4.3.5 Chemical analysis

Water samples were kept at 4 °C until analysis with online-solid phase extraction-liquid chromatography-tandem mass spectrometry SPE-HPLC-MS-MS. In brief, 18 mL filtrated water sample (cellulose acetate, 0.45 µm) were adjusted to pH 4 with acetate buffer and isotopic labeled pesticides were added as internal standards. Enrichment and elution of pesticides on a strata X column (2.1 mm I.D. x 20 mm l, Phenomenex) was performed with an online-system using methanol for elution (Freitas et al. 2004). Separation was carried out by an Xbridge C18 column (2 mm I.D. x 50 mm L, Waters) at room temperature with a flow rate of 200 µL/min. Water with 0.1 % formic acid and methanol with 0.1 % formic acid served as mobile phase using a gradient of 40 to 90 % methanol within 20 min. MS-MS detection (TSQ Quantum, Thermo Electrons, San José, CA, USA) was performed in the selected reaction mode (SRM) with positive electrospray ionization using a source voltage of 4.5 kV, ion transfer capillary temperature of 350 °C, sheath gas flow of 0.6 L/min and auxiliary gas flow of 1.5 L/min. The observed transitions of the pesticides and their internal standards are reported elsewhere (Stoob et al. 2005).

4.3.6 Taxonomic determination

For the taxonomic determination, aliquots of the fixed phytoplankton samples were taken and cells were allowed to settle in a sedimentation chamber for at least 24 h. Quantitative evaluation was done according to the method of Utermöhl described in Schwoerbel (2005) by using an inverted microscope (Diavert, Leitz, Oberkochen, Germany) with a 400 x magnification. Taxa were identified to species or to the lowest possible taxonomic level (Ettl et al. 1978-1991, Huber-Pestalozzi 1950-83). Cells counts were expressed as individuals/mL for each taxon.

4.3.7 Data analysis and statistical analysis

The dissipation of the herbicides in the post-treatment period was described by calculating decay rates and half-life periods according to first order kinetics. Phytoplankton data were summarized by determining total number of algae cells (total phytoplankton abundance), absolute and relative abundances of single algal groups and species and by determining the number of different taxa found per sample. On the taxon level, mean abundances per treatment level (single herbicide or mixture) were tested for significant differences to each other using Duncan's multiple t-test. The abundance data were log-transformed before the analysis to achieve normal distribution and variance homogeneity. The Shannon index (Boyle et al. 1990, Shannon and Weaver 1949) considering species number and distribution of individuals over species was used as a measure for the diversity of the phytoplankton community in a sample. In order to analyze in more detail the effects on the community structure principal component analysis (PCA) was used to create ordination diagrams per sampling date where the distance between the sample score represents the dissimilarity of the community in these samples. Only log-transformed abundances were used for the ordinations. Statistical analysis was performed with the Community Analysis computer program (developed by Hommen, 2007, an earlier version is described in Hommen et al. 1994) except for the PCAs which were performed using the CANOCO 4.5 (Biometrics Plant Research International, Wageningen, The Netherlands).

4.4 Results

4.4.1 Chemical analysis

During the five-week period of constant exposure (d 0 – 34) averaged atrazine, isoproturon, and diuron concentrations calculated as time weighted average were in the range of the respective target concentration ± 20 % in all mesocosms for all herbicides. At the end of the experiment in October (day 173, approximately 5 months after the end of the constant exposure period), atrazine concentrations in the single treatment were still high with 42.4 ± 2.4 % of the HC_{30} (n = 3) while diuron (11.6 ± 1.3 % of HC_{30}, n = 3) and isoproturon (8.4 ± 3.3 % of HC_{30}, n = 3) had nearly completely dissipated. In the mixture treatment atrazine, isoproturon, and diuron concentrations accounted for 13.5 ± 0.8 %, 1.7 ± 0.6 % and 3.1 ± 0.3 % of the respective HC_{30} at the end of the study. The difference in the dissipation of the compounds is also indicated in the half lives ($t_{1/2}$) of the herbicides since isoproturon ($t_{1/2}$ = 35 d, r^2 = 0.80) and diuron ($t_{1/2}$ = 43 d, r^2 = 0.97) dissipated 2 to 3 times faster than atrazine ($t_{1/2}$ = 107 d, r^2 = 0.92) (Knauert et al. 2008).

4.4.2 Total abundance

Prior to the first application (day -8) total abundance in all ponds was similar (one way ANOVA; p>0.05) (data not shown). During the period of constant exposure, i.e. in May, mean total abundance reached approximately 6000 individuals/mL in the control treatment (Fig. 4.2). In the atrazine, isoproturon, diuron as well as in the mixture treatment total abundance was similarly reduced by a factor of approximately three (Fig. 4.2) resulting in 1500 (isoproturon, diuron, mixture) and 2500 (atrazine) individuals/mL on day 26.

Abundances determined for the treated communities were not significantly different to those of the control from day 40 onwards with exceptions on day 110 and 173 (Fig. 4.2). At the end of the experiment in October (day 173) the total abundance in all treatments ranged between 3000 and 5000 individuals/mL and was significantly lowered only in the diuron treatment (Fig. 4.2).

4.4.3 Composition of phytoplankton assemblage

In summary, a total number of 147 different algal species in 150 phytoplankton samples were taxonomically identified at least to genus level. These taxa belonged to the following ten classes: Chlorophyceae, Prasinophyceae, Zygnematophyceae, Bacillariophyceae, Chrysophyceae, Xanthophyceae, Cryptophyceae and Cyanophyceae. The Chlorophyceae were the most dominant algal class comprising 45 % of all individuals averaged over all the 150 samples followed by the Chrysophyceae (23 %), Cryptophyceae (13 %) and Cyanophyceae (12 %). Overall, 21 species belonging to these four algal classes made up more than 90 % of all individuals. The most abundant species were *Choriocystis* cf. *minor* (17 % dominance), *Chromulina* cf. *obconica* (14 %), *Oocystis naegelii* (9 %), *Monoraphidium circinale* (8 %), *Chroococcus* sp. (7 %) and *Chroomonas acuta* (6 %).

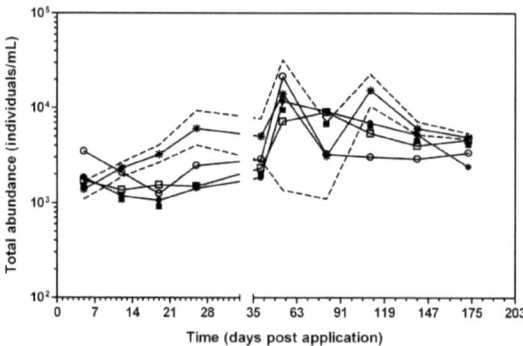

Fig. 4.2 Total abundance (individuals/mL) determined in the control (*), atrazine (○), diuron (●), isoproturon (□), and mixture (■) treatment over the entire course of the experiment. Dashed lines indicate the minimum and maximum values of the control community. The first part of the time axis corresponds to the period of constant exposure and each minor tick represents one exposure day. The second part of the time axis corresponds to the post-treatment period while minor ticks represent one exposure week.

4.4.4 Community structure and succession

On day 5, the phytoplankton community of the controls was mainly dominated by the Chrysophyceae and the Cryptophyceae (each 40 % of total abundance) (Fig. 4.3). Two to three weeks later (day 19 to 26), the Cryptophytes comprised up to 80 % of total abundance whereas Chrysophyceae reduced to 10 % only. In particular the abundance of *Chroomonas acuta* increased resulting in approximately 6000 individuals/mL (day 26) (Fig. 4.4). Later, at the end of June, the abundance of the Chlorophytes increased yielding more than 80 % of total phytoplankton due to a bloom of *Choriocystis* cf. *minor* and *Monoraphidium circinale* (Fig. 4.3&4.4). During August (day 82 to 110) Chrysophyceae and Cyanophyceae became the dominant algae comprising 80 % of the phytoplankton (Fig. 4.3). Among the Cyanophyceae and the Chrysophyceae, *Chroococcus* sp. and *Chromulina* cf. *obconica* became most abundant with approximately 9000 individuals/mL and 3000 individuals/mL (day 110). During September and October (day 140 to 173), the Chlorophyceae made up 60 % of total phytoplankton dominated by *Oocystis naegelii* with approximately 3000 individuals/mL (Fig. 4.3&4.4).

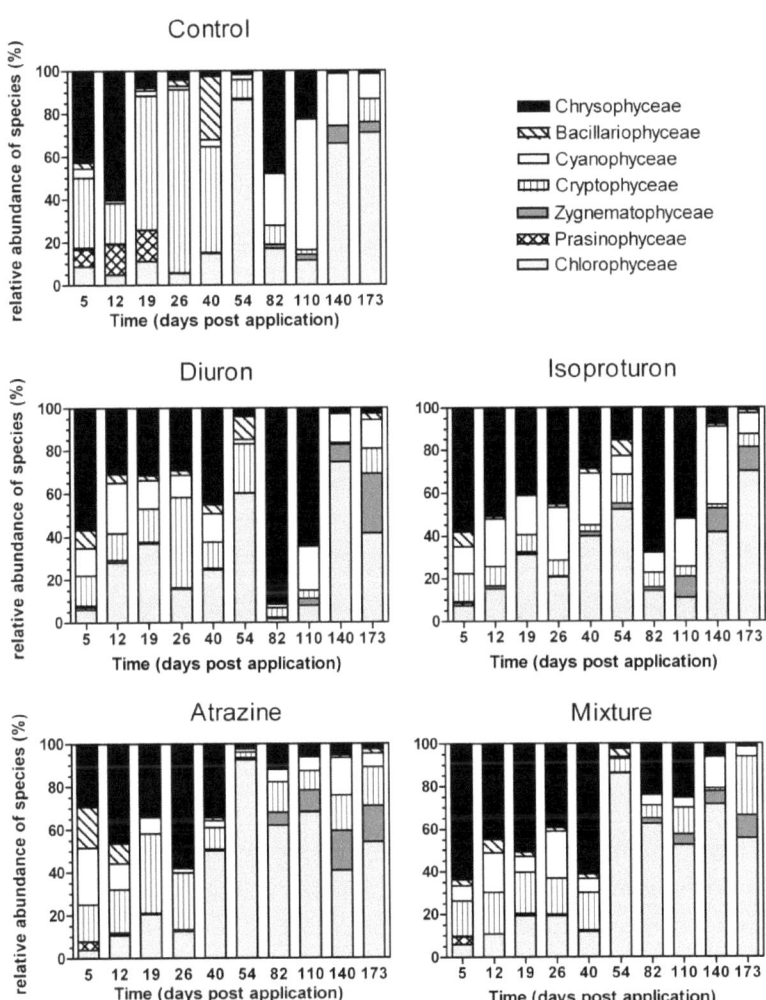

Fig. 4.3 Relative abundance of major algal classes in the control and the various treatments. Days 5 to 26 represent the period of constant exposure. Days 40 to 173 represent the post-treatment period.

Prior to (day -8) and short time after application (day 5), the species composition of the treated communities was not significantly different from that of the control communities (Fig. 4.5, Table 4.1). From day 12 to day 26, differences between the control and the treated mesocosms were observed. Species composition of phytoplankton in the various treatments however did not differ from each other with some minor exceptions on days 19 and 26 (Table 4.1). In contrast to the control mesocosms, Chrysophyceae, Chlorophyceae, and Cyanophyceae mainly dominated the treated communities from day 12 to day 26 (Fig. 4.3). In particular, *Chroomonas acuta* and *Cryptomonas erosa et ovata* were herbicide sensitive (Fig. 4.4). Due to differences in their susceptibility, the Cryptophytes were reduced to 20 % of the total abundance in the isoproturon and the mixture treatment compared to 40 % in the atrazine and diuron treatment (day 12 to 26) (Fig. 4.3&4.4). Abundance of the Prasinophyte *Nephroselmis* cf. *olivaceae* was found to be similarly reduced by the three herbicides and the mixture whereas growth of the Cryptophyte *Katablepharis ovalis* similarly increased in the various treatments compared to the control (day 5 to 19). Since *Ankyra judayi* was susceptible to atrazine and the mixture but not to isoproturon and diuron the Chlorophytes achieved lower relative abundances in the atrazine and mixture compared to the isoproturon and diuron treatments (Fig. 4.3).

During the post-treatment period, the phytoplankton communities treated with diuron and isoproturon developed similar to the control communities and were also similar to each other concerning their community composition from day 54 onwards with the exceptions of days 82 and 140 for the diuron treatment only (Fig. 4.5, Table 4.1). As observed for the control phytoplankton communities, the Chlorophyceae, in particular *Choriocystis* cf. *minor* and *Monoraphidium circinale*, dominated the communities in the diuron and isoproturon treatment on day 54 (Fig. 4.3). During August (day 82 to 110) the Chrysophyceae and Cyanophyceae became more abundant in the diuron and isoproturon treatment as it was also observed for the control community (Fig. 4.3). However the Chrysophyceae achieved higher and Cyanophyceae lower relative abundances due to enhanced and increased growth of *Chromulina* cf. *obconica* and *Chroococcus* sp. (Fig. 4.3&4.4). During September and October (day 140 to 173), the Chlorophyceae became the dominant group as similarly described for the control communities. In contrast to the control, *Choriocystis* cf. *minor* was the most abundant green alga in the diuron treatment on day 140 whereas *Oocystis naegelii* was found to be sensitive to diuron (Fig. 4.3&4.4).

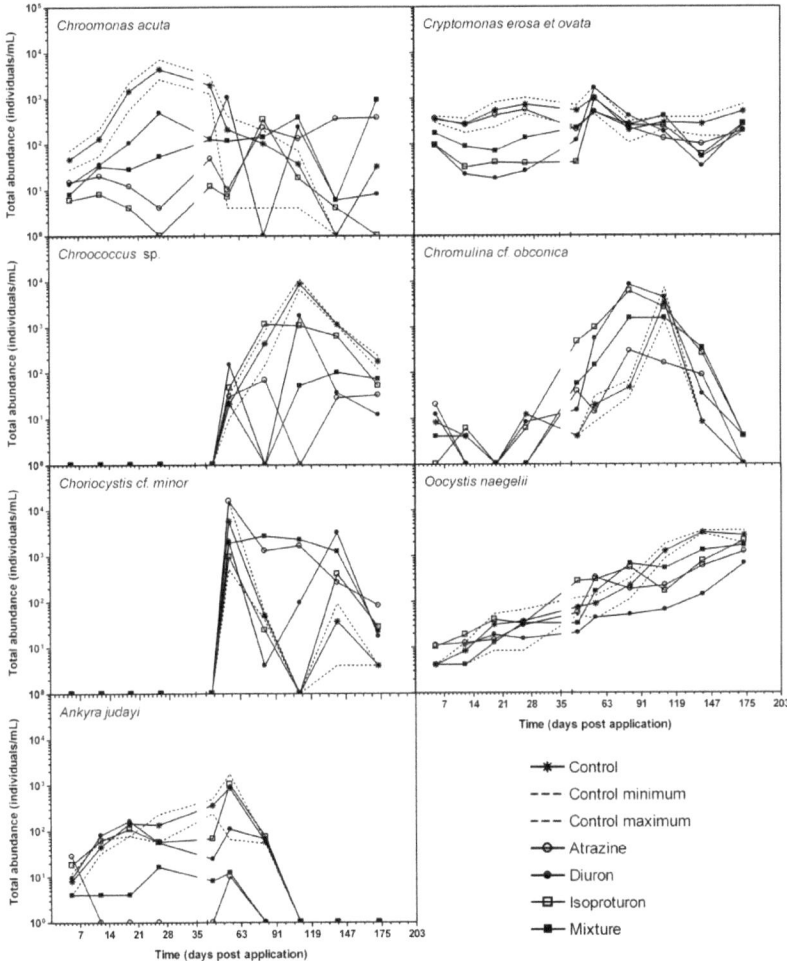

Fig. 4.4 Absolute abundances of the two Cryptophytes *Chroomonas acuta* and *Cryptomonas erosa et ovata*, the Cyanophyte *Chroococcus* sp., the Chrysophyte *Chromulina* cf. *obconica* and the three Chlorophytes *Choriocystis* cf. *minor*, *Oocystis naegelii* and *Ankyra judayi* in the different treatments. Data points localized on the x axis indicate that the abundance of the respective species was below the detection limit of microscopic analysis on these sampling days. The first part of the time axis corresponds to the period of constant exposure and each minor tick represents one exposure day. The second part of the time axis corresponds to the post-treatment period while minor ticks represent one exposure week.

Chapter 4 – Effects of PSII inhibitors and their mixture on phytoplankton succession

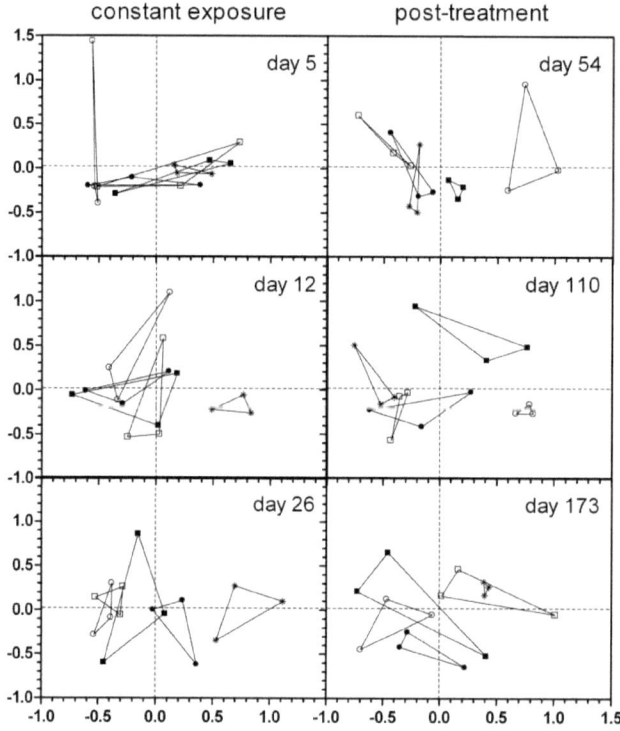

Fig. 4.5 Plots of principal component analysis (PCA) sample scores for the phytoplankton data from six different sampling days. Each data point indicates one replicate. * = control, ○ = atrazine, ● = diuron, □ = isoproturon and ■ = herbicide mixture.

In contrast to the diuron and isoproturon treated communities, atrazine and mixture treated communities remained different in species composition compared to the control in the post-treatment period until the end of the experiment (Fig. 4.5, Table 4.1). The Chlorophyceae were found to be the most dominant algal group in the atrazine and mixture treatment comprising 40 to 90 % of total abundance (Fig. 4.3). The bloom of the Cyanophyceae and Chrysophyceae observed in the control ponds during August was nearly completely suppressed in the atrazine and mixture treated ponds due to the sensitivity of *Chroococcus* sp. towards atrazine and the mixture (day 82 to 140) and *Chromulina* cf. *obconica* towards atrazine (day 110) (Fig 4.3&4.4). In contrast, growth of *Choriocystis* cf. *minor* was significantly enhanced in the atrazine and the mixture treatment compared to the control (days 82 to 110) (Fig. 4.4).

Table 4.1 PCA results were statistically analyzed by Duncan's multiple t-test (p<0.05). Crosses indicate statistically significant differences between the phytoplankton community structure of the two treatments that are compared. con = control, atra = atrazine, iso = isoproturon, diu = diuron, mix = mixture.

	pre-treatment	constant exposure				post-treatment					
day	-8	5	12	19	26	40	54	82	110	140	173
con vs. atra	-	-	x	x	x	x	x	x	x	x	x
con vs. diu	-	-	x	x	x	x	-	x	-	x	-
con vs. iso	-	-	x	x	x	x	-	-	-	-	-
con vs. mix	-	-	x	x	x	x	x	x	x	x	-
atra vs. diu	-	-	-	x	x	x	x	-	x	-	-
atra vs. iso	-	-	-	-	-	-	x	-	x	x	x
atra vs. mix	-	-	-	-	-	-	x	-	-	x	-
diu vs. iso	-	-	-	x	x	x	-	-	-	x	-
diu vs. mix	-	-	-	-	-	x	x	-	-	-	-
iso vs. mix	-	-	-	-	-	-	x	x	x	-	-

4.4.5 Number of different taxa and Shannon diversity

Prior to exposure (day -8), the averaged number of different taxa in the ponds accounted for 30 ± 3 (data not shown). On day 5, the control communities comprised 23 ± 6 different species (Fig. 4.6). The number of taxa in the single herbicide as well as the mixture treatments was comparable to the control, during the period of constant exposure with few exceptions on day 5 and 19. In the post-treatment period, the number of different species at the various sampling dates was still comparable in the treatments and the control ranging between 16 and 35 different species.

Prior to (day -8; data not shown) and directly after application (day 5), the Shannon indices of the control and the treated communities were similar (Fig. 4.6). However, in the subsequent three exposure weeks the Shannon index of the control decreased by a factor of two indicating an unbalanced distribution of the organisms among the various taxonomic groups due to the dominance of a few species namely the Cryptophyceae *Chroomonas acuta*, *Cryptomonas erosa et ovata* and *Katablepharis ovalis* (Fig. 4.4).

In contrast, Shannon index of the phytoplankton communities in the single herbicide and the mixture treatments remained similarly constant (Fig. 4.6). Considering the comparable number of different species in the various treatments, the high Shannon index determined for the treated communities indicated a more homogenous distribution of individuals. During the post-treatment period, in general, no important differences in the Shannon index were observed.

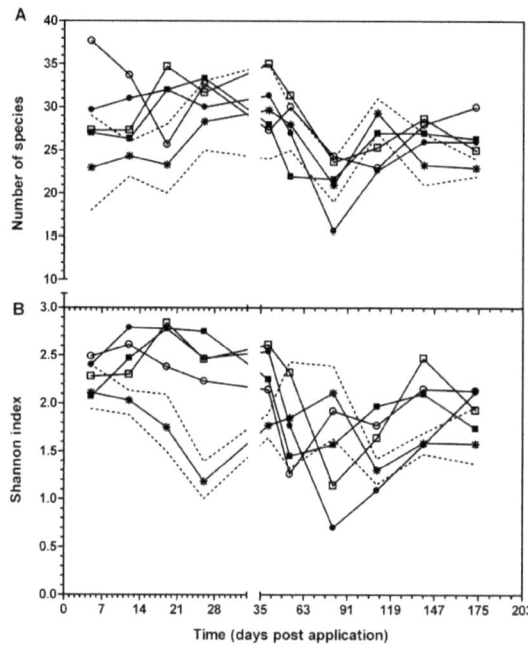

Fig. 4.6 Diversity of control and exposed phytoplankton communities. Total number of algal species (A) and Shannon index (B) determined for the control (*), atrazine (○), diuron (●), isoproturon (□) and mixture (■) treatment are presented. Dashed lines indicate the minimum and maximum values of the control community. The first part of the time axis corresponds to the period of constant exposure and each minor tick represents one exposure day. The second part of the time axis corresponds to the post-treatment period while minor ticks represent one exposure week.

4.5 Discussion

In this freshwater outdoor mesocosm study, the atrazine, isoproturon, and diuron treatments were demonstrated to significantly influence phytoplankton community parameters such as abundance, diversity, species composition and community structure due to the sensitivity of a few algal species. The vast majority of the taxonomically identified algal species in this mesocosm experiment remained unaffected by the three PSII herbicides and their mixture.

During the period of constant concentrations, the HC_{30} of atrazine, isoproturon, and diuron were shown to be equitoxic concerning their effects on total abundance and diversity of phytoplankton despite the different sensitivities of single algal species towards the herbicides. Moreover, we observed that a mixture comprising 1/3 of the HC_{30} of each herbicide elicited effects on total abundance and diversity of phytoplankton comparable to the single active substances indicating that the mixture toxicity concept of concentration addition applies also for such higher levels of biological organization in the field situation.

In the long term, species composition in the various treatments developed different from each other due to differences in the sensitivity of a few algal species towards the herbicides as indicated by unaffected diversity and species numbers. Recovery with respect to species composition was observed already two weeks after constant exposure for the diuron and isoproturon treated phytoplankton and can be explained by the fast dissipation of the two phenylurea herbicides. In contrast, recovery of the mixture and atrazine treated communities were not observed until the end of the experiment. This might be explained by the still high exposure to atrazine in these treatments as a consequence of the slow dissipation of atrazine in the water column and by differences in the succession of the phytoplankton in these mesocosms.

Furthermore, species sensitivities detected in the mesocosm experiment were compared to the laboratory data underlying the SSDs curves that were used to derive the HC_{30}. SSD curves were based on EC_{50} growth data of some green algae and diatom species. Since taxa of both groups were in general randomly distributed in the SSD neither group was thus expected to be more sensitive towards atrazine, isoproturon, and diuron. Consistently, neither of these both algal groups but also none of the other identified groups was predominantly found to be most sensitive towards one of the herbicides in this study. In fact, PSII herbicide sensitive algal species were detected throughout different algal classes such as the Cryptophyceae, Prasinophyceae, Cyanophyceae, Chlorophyceae, and Chrysophyceae. But, in contrast to the assumed sensitivity of the diatoms regarding the SSD data, species of this group remained completely unaffected due to herbicide stress in this study. This might be explained by the fact that none of the diatom species used to construct the SSD was actually abundant in our phytoplankton communities. The green alga *Scenedesmus quadricauda* was the only species present in the SSD and abundant in our natural phytoplankton community. *Scenedesmus quadricauda* was expected to be affected by the chosen atrazine concentration since the EC_{50} accounted for 85 µg/L. However, no adverse effects on growth of *Scenedesmus quadricauda* were observed indicating that this species was less sensitive to atrazine in the natural community than in a laboratory monoculture.

Comparing our observations to results obtained from another mesocosm study performed by Grünwald, a similar susceptibility of certain algae, such as *Chroomonas acuta, Cryptomonas erosa et ovata,* and *Nephroselmis* cf. *olivaceae,* towards isoproturon as described in this study was reported (Grünwald 2003). Growth of *Cryptomonas erosa* remained unaffected upon exposure to 100 µg atrazine/L in aquatic enclosures placed in Lake St. George Canada (Hamilton et al. 1988) which is also in line with the observations of the present study. However, the green alga *Chlorella vulgaris* was found to be tolerant to 70 µg atrazine/L in this study whereas Bérard et al. (1999) detected significant fewer numbers of *Chlorella vulgaris* in phytoplankton communities sampled from Lake Geneva that were exposed to 10 µg atrazine/L in indoor microcosms. Differences in sensitivities of single phytoplankton species might be explained by varying environmental parameters such as light, temperature or nutrient availability and by the species-species interactions of the actually affected community (Guasch et al. 1998, 2007). Certain species such as the green algae *Choriocystis* cf. *minor,* the Chrysophyte *Chromulina* cf. *obconica,* and the Cryptophyte *Katablepharis ovalis* were enhanced in growth possibly as a consequence of reduced competition for nutrients and light.

Overall, our results confirmed the applicability of the concept of concentration addition for PSII inhibitors when considering their effects on abundance and diversity of a natural algal community under environmental conditions. These results can thus contribute to the current discussion concerning the incorporation of mixture toxicity in the regulation of surface water quality to adequately protect aquatic communities from adverse pesticide impact and to guarantee a sustained management of the aquatic ecosystems.

4.6 References

Arrhenius, A., Gronvall, F., Scholze, M., Backhaus, T., Blanck, H., 2004. Predictability of the mixture toxicity of 12 similarly acting congeneric inhibitors of photosystem II in marine periphyton and epipsammon communities. *Aquatic Toxicology* 68, 351-367.

Arrhenius, A., Backhaus, T., Gronvall, F., Junghans, M., Scholze, M., Blanck, H., 2006. Effects of three antifouling agents on algal communities and algal reproduction: Mixture toxicity studies with TBT, Irgarol, and Sea-Nine. *Archives of Environmental Contamination and Toxicology* 50, 335-345.

Babut, M., Bonnet, C., Bray, M., Flammarion, P., Garric, J., Golaszewski, G., 2003. Developing environmental quality standards for various pesticides and priority pollutants for French freshwaters. *Journal of Environmental Management* 69, 139-147.

Backhaus, T., Arrhenius, A., Blanck, H., 2004a. Toxicity of a mixture of dissimilarly acting substances to natural algal communities: Predictive power and limitations of independent action and concentration addition. *Environmental Science & Technology* 38, 6363-6370.

Backhaus, T., Faust, M., Scholze, M., Gramatica, P., Vighi, M., Grimme, L.H., 2004b. Joint algal toxicity of phenylurea herbicides is equally predictable by concentration addition and independent action. *Environmental Toxicology and Chemistry* 23, 258-264.

Bérard, A., Pelte, T., Druart, J.C., 1999. Seasonal variations in the sensitivity of Lake Geneva phytoplankton community structure to atrazine. *Archiv für Hydrobiologie* 145, 277-295.

Bérard, A., Dorigo, U., Mercier, I., Becker-van Slooten, K., Grandjean, D., Leboulanger, C., 2003. Comparison of the ecotoxicological impact of the triazines Irgarol 1051 and atrazine on microalgal cultures and natural microalgal communities in Lake Geneva. *Chemosphere* 53, 935-944.

Blanchoud, H., Farrugia, F., Mouchel, J.M., 2004. Pesticide uses and transfers in urbanised catchments. *Chemosphere* 55, 905-913.

Boyle, T.P., Smilie, G.M., Anderson, J.C., Beeson, D.R., 1990. A sensitive analysis of nine diversity and seven similarity indices. *Research Journal Water Pollution Control Federation* 62, 749-762.

Chèvre, N., Loeppe, C., Singer, H., Stamm, C., Fenner, K., Escher, B.I., 2006. Including mixtures in the determination of water quality criteria for herbicides in surface water. *Environmental Science & Technology* 40, 426-435.

Dorigo, U., Bourrain, X., Bérard, A., Leboulanger, C., 2004. Seasonal changes in the sensitivity of river microalgae to atrazine and isoproturon along a contamination gradient. *Science of the Total Environment* 318, 101-114.

Drost, W., Backhaus, T., Vassilakaki, M., Grimme, L.H., 2003. Mixture toxicity of s-triazines to *Lemna minor* under conditions of simultaneous and sequential exposure. *Fresenius Environmental Bulletin* 12, 601-607.

Ettl, D., Gerloff, J., Heynig, M., Mollenhauer, D., 1978-1991. *Die Süßwasserflora von Mitteleuropa.* Gustav Fischer Verlag, Stuttgart, Germany.

Faust, M., Altenburger, R., Backhaus, T., Blanck, H., Boedeker, W., Gramatica, P., Hamer, V., Scholze, M., Vighi, M., Grimme, L.H., 2001. Predicting the joint algal toxicity of multi-component s-triazine mixtures at low-effect concentrations of individual toxicants. *Aquatic Toxicology* 56, 13-32.

Field, J.A., Reed, R.L., Sawyer, T.E., Griffith, S.M., Wigington, P.J., 2003. Diuron occurrence and distribution in soil and surface and ground water associated with grass seed production. *Journal of Environmental Quality* 32, 171-179.

Freitas, L.G., Goetz, C.W., Ruff, M., Singer, H.P., Müller, S.R., 2004. Quantification of the new triketone herbicides, sulcotrione and mesotrione, and other important herbicides and metabolites, at the ng/l level in surface waters using liquid chromatography-tandem mass spectrometry. *Journal of Chromatography A* 1028, 277-286.

Graymore, M., Stagnitti, F., Allinson, G., 2001. Impacts of atrazine in aquatic ecosystems. *Environment International* 26, 483-495.

Grünwald, H. M., 2003. *Effects of a pesticide mixture on plankton in freshwater mesocosms - from single substance studies to combination impacts.* Dissertation TU München.

Guasch, H., Ivorra, N., Lehmann, V., Paulsson, M., Real, M., Sabater, S., 1998. Community composition and sensitivity of periphyton to atrazine in flowing waters: the role of environmental factors. *Journal of Applied Phycology* 10, 203-213.

Guasch, H., Lehmann, V., van Beusekom, B., Sabater, S., Admiraal, W., 2007. Influence of phosphate on the response of periphyton to atrazine exposure. *Archives of Environmental Contamination and Toxicology* 52, 32-37.

Gustavson, K., Mohlenberg, F., Schluter, L., 2003. Effects of exposure duration of herbicides on natural stream periphyton communities and recovery. *Archives of Environmental Contamination and Toxicology* 45, 48-58.

Hamilton, P.B., Jackson, G.S., Kaushik, N.K., Solomon, K.R., Stephenson, G.L., 1988. The Impact of 2 Applications of Atrazine on the Plankton Communities of Insitu Enclosures. *Aquatic Toxicology* 13, 123-139.

Hartgers, E.M., Aalderink, G.H.R., Van den Brink, P.J., Gylstra, R., Wiegman, J.W.F., Brock, T.C.M., 1998. Ecotoxicological treshold levels of a mixture of herbicides (atrazine, diuron and metolachlor) in freshwater microcosms. *Aquatic Ecology* 32, 135-152.

Hommen, U., Veith, D., Dülmer, U., 1994. *A computer program to evaluate plankton data from freshwater field tests*, in: Freshwater field tests for hazard assessment of chemicals, eds. Hill, I.R., Heimbach, F., Leeuwangh, P., Matthiessen, P.,. Lewis Publishers, Boca Raton, FL, USA, pp. 503-514.

Huber, W., 1993. Ecotoxicological Relevance of Atrazine in Aquatic Systems. *Environmental Toxicology and Chemistry* 12, 1865-1881.

Huber-Pestalozzi, G., 1950-1983. *Das Plankton des Süßwassers. Die Binnengewässer.* E. Schweizerbart'sche Verlagsbuchhandlung, Stuttgart, Germany.

Irace-Guigand, S., Aaron, J.J., Scribe, P., Barcelo, D., 2004. A comparison of the environmental impact of pesticide multiresidues and their occurrence in river waters surveyed by liquid chromatography coupled in tandem with UV diode array detection and mass spectrometry. *Chemosphere* 55, 973-981.

Jahnel, J., Zwiener, C., Gremm, T.J., Abbt-Braun, G., Frimmel, F.H., Kussatz, C., Schudoma, D., Rocker, W., 2001. Quality targets for pesticides and other pollutants in surface waters. *Acta Hydrochimica Et Hydrobiologica* 29, 246-253.

Junghans, M., Backhaus, T., Faust, M., Scholze, M., Grimme, L.H., 2003. Predictability of combined effects of eight chloroacetanilide herbicides on algal reproduction. *Pest Management Science* 59, 1101-1110.

Kirby, M.F., Sheahan, D.A., 1994. Effects of Atrazine, Isoproturon, and Mecoprop on the Macrophyte *Lemna minor* and the Alga *Scenedesmus subspicatus. Bulletin of Environmental Contamination and Toxicology* 53, 120-126.

Knauert, S., Escher, B., Singer, H., Hollender, J., Knauer, K., 2008. Mixture toxicity of three photosystem II inhibitors (atrazine, isoproturon, and diuron) towards photosynthesis of freshwater phytoplankton studied in outdoor mesocosms. *Environmental Science & Technology*

Kreuger, J., 1998. Pesticides in stream water within an agricultural catchment in southern Sweden, 1990-1996. *The Science of the Total Environment* 216, 227-251.

Nitschke, L., Schüssler, W., 1998. Surface water pollution by herbicides from effluents of waste water treatment plants. *Chemosphere* 36, 35-41.

Nowell, L., Resek, E., 1994. *National standards and guidelines for pesticides in water, sediment, and aquatic organisms: application to water-quality assessment*, in: Reviews of environmental contamination and toxicology Vol. 140, ed. Ware, G.W.,. Springer Verlag, New York, USA, pp. 1-164.

Peres, F., Florin, D., Grollier, T., Feurtet-Mazel, A., Coste, M., Ribeyre, F., Ricard, M., Boudou, A., 1996. Effects of the phenylurea herbicide isoproturon on periphytic diatom communities in freshwater indoor microcosms. *Environmental Pollution* 94, 141-152.

Schwarzenbach, R.P., Escher, B.I., Fenner, K., Hofstetter, T.B., Johnson, C.A., von Gunten, U., Wehrli, B., 2006. The challenge of micropollutants in aquatic systems. *Science* 313, 1072-1077.

Schwoerbel, J., 2005. *Einführung in die Limnologie*. 9th edition, Spektrum Akademischer Verlag, Heidelberg, Germany.

Shannon, C.E., Weaver, W., 1949. *The mathematical theory of communication*. University of Illinois Press, Urbana, IL, USA.

Solomon, K.R., Baker, D.B., Richards, R.P., Dixon, D.R., Klaine, S.J., LaPoint, T.W., Kendall, R.J., Weisskopf, C.P., Giddings, J.M., Giesy, J.P., Hall, L.W., Williams, W.M., 1996. Ecological risk assessment of atrazine in North American surface waters. *Environmental Toxicology and Chemistry* 15, 31-74.

Stephan, C., Mount, D., Hansen, D., Gentile, J., Chapman, G., Brungs, W., 1985. *Guidelines for deriving numerical national water quality criteria for the protection of aquatic organisms and their uses*; NTIS no PB85-227049; US Environmental Protection Agency, Washington DC.

Stoob, K., Singer, H.P., Goetz, C.W., Ruff, M., Müller, S.R., 2005. Fully automated online solid phase extraction coupled directly to liquid chromatography-tandem mass spectrometry -

Quantification of sulfonamide antibiotics, neutral and acidic pesticides at low concentrations in surface waters. *Journal of Chromatography A* 1097, 138-147.

Teisseire, H., Couderchet, M., Vernet, G., 1999. Phytotoxicity of diuron alone and in combination with copper or folpet on duckweed (*Lemna minor*). *Environmental Pollution* 106, 39-45.

Traunspurger, W., Schäfer, H., Remde, A., 1996. Comparative investigation on the effect of a herbicide on aquatic organisms in single species tests and aquatic microcosms. *Chemosphere* 33, 1129-1141.

Trebst, A., 1987. The 3-dimensional structure of the herbicide binding niche on the reaction center polypeptides of photosystem II. *Zeitschrift für Naturforschung C - Journal of Biosciences* 42, 742-750.

Warne. M.S.J., 2002. *A review of the ecotoxicity of mixtures, approaches to, and recommendations for, their management*, in: Proceedings of the Fifth National Workshop on the Assessment of Site Contamination, eds. Langley, A., Gilbey, M., Kennedy, B., NEPC Service Corporation, Adelaide, Australia, pp 253-276.

Zabel, T., Cole, S.M., 1999. The derivation of environmental quality standards for the protection of aquatic life in the UK. *The Chartered Institution of Water and Environmental Management Journal* 13, 436-440.

Chapter 5

Phytotoxicity of atrazine, isoproturon, and diuron to submersed macrophytes in outdoor mesocosms

5.1 Abstract

The three submersed macrophytes *Elodea canadensis* Michx., *Myriophyllum spicatum* L., and *Potamogeton lucens* L. were constantly exposed over a five-week period to environmental relevant concentrations of the photosystem II inhibiting herbicides atrazine, isoproturon, diuron, and to their mixture in outdoor mesocosms. Effects on the three macrophytes were evaluated investigating photosynthetic efficiency. The method used to determine photosynthetic efficiency was based on *in vivo* chlorophyll fluorescence measurements. In addition, effects on *M. spicatum* and *E. canadensis* were assessed by calculating relative growth rates based on total length measurements. Adverse effects on the photosynthetic efficiency of the three macrophytes were observed on days 2 and 5 after application. Considering photosynthetic efficiency, *M. spicatum* was found to be the most sensitive macrophyte whereas *E. canadensis* and *P. lucens* were less sensitive to atrazine, diuron, and the mixture and insensitive to isoproturon. Growth of *E. canadensis* and *M. spicatum* was not reduced which indicates that the herbicide exposure did not impair plant development. Photosynthetic efficiency of *M. spicatum* was similar affected in the single as well as in the mixture treatments indicating that the herbicides acted concentration additive. Concerning *E. canadensis* and *P. lucens,* it was difficult to draw any conclusions on mixture toxicity from our results. Since mixtures are the common exposure scenario in the environment, we recommend taking mixture toxicity into account while monitoring and regulating surface water quality with respect to herbicide contamination.

5.2 Introduction

Aquatic plants are of great importance in stabilizing function and structure of freshwater ecosystems. As primary producers, macrophytes play a major role in the aquatic carbon and nutrient cycling. They provide food as well as habitats for aquatic communities. Especially in oligotrophic ponds and lakes as well as in streams and wetland communities submersed floating and emergent macrophytes are essential to harbor diverse animal communities (Wetzel 2001). Especially rooted macrophytes greatly improve water quality and clarity by damping wave activity, stabilizing sediments and therefore reducing shoreline erosion (Carpenter and Lodge 1986, Knauer 1993). Regardless the causes, any significant reduction in macrophytes can be expected to have a strong impact on the whole ecosystem (Lewis 1995).

The loss of native plant vegetation in freshwater systems throughout the world has become a great concern (Sand-Jensen et al. 2000, Körner 2002). One of the main factors causing such a decline in plant populations is, besides eutrophication, the elevated concentrations of herbicides in the aquatic environment. Herbicides are used for crop protection in agriculture, for green space management in urban areas, for plant management in wetlands and lakes and as antifouling paints. There are several ways, herbicides may enter aquatic ecosystems, e.g., spray drift, runoff, drainage, waste water discharge or accidental spills. Traces of single herbicides and herbicide mixtures have frequently been detected in aquatic ecosystems throughout the world. The three herbicides atrazine, isoproturon, and diuron are often determined in surface waters as single substances but also in mixture (Kreuger 1998, Gilliom et al. 1999, Gfrerer et al. 2002, Huang et al. 2004, Irace-Guigand et al. 2004). Since atrazine was banned in most European countries, concentrations in surface waters are generally below 1 µg/L atrazine (Garmouma et al. 1998, Graymore et al.2001). However, atrazine is still widely used in North America and Australia and concentrations in streams and rivers reach up to 150 µg/L in surface waters (Graymore et al. 2001, Key et al 2007). For isoproturon, concentrations of 17 µg/l were measured in agricultural run-off waters (Kirby and Sheahan 1994). Maximum diuron concentrations corresponding to 10 µg/L (Blanchoud et al. 2004) and 30 µg/L (Field et al. 2003) have been found in French and North American surface waters, respectively. Atrazine, isoproturon, and diuron belong to the urea/triazine type family of photosystem II (PSII) inhibitors and exert the same mode of toxic action. They interfere with the electron transport of PSII by competing with plastoquinon for binding to the D1 protein in the thylakoid membrane of plastids (Trebst 1987). The influence of environmental relevant concentrations on the development of macrophytes has been intensively studied for atrazine (see for review: Huber 1993, Solomon et al. 1996). However, little is known about the effects of isoproturon and diuron on macrophytes in their natural habitats (Lambert et al. 2006).

To assess effects of herbicides on macrophytes, plant growth based on biomass or root/shoot length has been investigated in laboratory as well as in field studies (Knauer et al. 2006). Other common endpoints were total chlorophyll content (Teisseire et al. 1999) and photosynthesis efficiency measured via chlorophyll fluorescence (Snel et al. 1998, Marwood et al. 2001, Fai et al. 2007).

Few studies reported on the potential threat to macrophytes exposed to herbicide mixtures in aquatic model ecosystems (Fairchild et al. 1994, Lytle and Lytle 2002, Lytle and Lytle 2005).

However, these studies neither reported on differences in species sensitivities nor on joint toxicity of herbicide mixtures.

In this context, the present study assessed the effects of environmental relevant concentrations of atrazine, isoproturon, diuron, and of their mixture on photosynthetic efficiency and growth of the three submersed macrophytes *Elodea canadensis, Myriophyllum spicatum,* and *Potamogeton lucens* in an outdoor mesocosm experiment. Our first objective was to determine species-specific sensitivities of the three macrophytes to the single and mixture herbicide treatments. A second objective was to compare effects from the single herbicide exposure to the mixture treatment with respect to additive phytotoxicity.

5.3   Material and methods

5.3.1   Outdoor mesocosm test site

The mesocosm test site of Syngenta Crop Protection AG was located in CH-8260 Stein, Switzerland (47°33'10" N, 7°57'47" E, 300 m a.s.l.). Each mesocosm had a volume of approximately 10 m^3, the surface had a diameter of 3 m, and a sediment and water layer of 15 and 130 cm, respectively (Knauer et al. 2005). To minimize rapid temperature fluctuations, the tanks were buried in the ground. Water circulation between the mesocosms and a nearby man-made supply pond (500 m^3) was started in March, when all ponds were thawed, and lasted for approximately six weeks. Before application, the circulation was stopped and the mesocosms were closed. Plankton, macroinvertebrates and other organisms were introduced into the mesocosms along with the water and sediment from the supply pond and via aerial colonization during the year. Macrophyte populations of *Elodea canadensis* developed from plants in the sediment.

5.3.2   Chemicals and exposure concentrations

Atrazine (2-chloro-4-ethylamino-6-isopropyl-amino-*s*-triazine) (99 % purity) and diuron (3-(3,4-dichlorophenyl)-1,1-dimethylurea) (purity 98.4 %) were provided by Syngenta (Basel, Switzerland) and DuPont Crop Protection (Newark, DE, USA), respectively. Isoproturon [3-(4-isopropylphenyl)-1,1-dimethylurea] (analytical standard) was purchased from Sigma-Aldrich (Buchs, Switzerland). Herbicides were dissolved in methanol (p.a., Merck, Darmstadt, Germany) prior to application whereby the final solvent concentration in the ponds was negligible (< 0.002 %). Atrazine, isoproturon and diuron concentrations in the single treatments corresponded to 70 µg/L = 325 nM, 14 µg/L = 66 nM and 5 µg/L = 21 nM, respectively. Each herbicide contributed to the herbicide mixture with 1/3 of the concentrations used in the single treatments taking the concept of concentration addition into account. Generally, herbicide concentrations were chosen because they were in the range of measured environmental concentrations. Further, the toxic potential of the single herbicides and their mixture were expected to be comparable since these concentrations correspond to the HC_{30} derived from spieces sensitivity distributions of the herbicides as already presented in Knauert et al. (2008).

5.3.3 Application and sampling

Prior to application, mesocosms were randomly assigned to the different treatments. Each treatment and the control were replicated three times. In total 15 mesocosms were investigated. Exposure started on 3 May 2006 (day 0). After the first application (day 0), herbicide concentrations in the single and mixture treatments were kept rather constant over a period of five weeks, i.e. the herbicides were supplemented to the ponds if necessary to maintain target concentrations at approximately ± 20 %. The herbicides were applied with a sprayer on the water surface of the mesocosms. The water column was than stirred for approximately 5 min with a polyethylene tube to achieve a homogenous distribution of the test compound. Separated sampling equipments such as tubes, buckets etc. were used for each individual treatment to avoid cross contamination.

Sampling for chemical water analysis to control herbicide concentrations started on the day of application. For chemical analysis depth-integrated water samples were taken from the four quadrants of each mesocosm using a polyethylene tube, 120 cm long and 4.5 cm in diameter. Samples were taken by lowering this tube close to the mesocosm sediment surface to avoid sample contamination due to sediment perturbation. For chemical analysis, samples were transported directly after sampling on ice to the laboratory and measured immediately to decide on redosing. Concentrations to be re-applied were calculated based on these measurements.

5.3.4 Chemical analysis

Analysis was performed with online-solid phase extraction-liquid chromatography-tandem mass spectrometry SPE-HPLC-MS-MS. In brief, 18 ml filtrated water sample (cellulose acetate, 0.45 μm) were adjusted to pH 4 with acetate buffer. Isotopic labeled pesticides were added as internal standards. Enrichment and elution of pesticides on a strata X column (2.1 mm I.D. x 20 mm I, Phenomenex) was performed with an online-system using methanol for elution (Freitas et al. 2004). Separation was achieved by an Xbridge C18 column (2 mm I.D. x 50 mm L, Waters) at room temperature with a flow rate of 200 μl/min. The mobile phase was water with 0.1 % formic acid and methanol with 0.1 % formic acid, using a gradient of 40 % to 90 % methanol within 20 min. MS-MS detection (TSQ Quantum, Thermo Electrons, San Jose, CA, USA) was carried out in the selected reaction mode (SRM) with positive electrospray ionization using a source voltage of 4.5 kV, ion transfer capillary temperature of 350 °C, sheath gas flow of 0.6 l/min and auxiliary gas flow of 1.5 l/min. The monitored transitions of the pesticides and their internal standards are reported elsewhere (Freitas et al. 2004). Limits of quantification using signal-to-noise ratio >10 were 3 ng/L.

5.3.5 Physical/chemical water parameters and nutrients

Temperature, pH, dissolved oxygen and conductivity (WTW, Effretikon, Switzerland) were measured at a water depth of approximately 20 cm in the morning at the same time in weekly intervals. In addition, a 50 ml water sample was taken prior to application on day -23 and during exposure on day 16 for the determination of nitrate and phosphate. Water samples were stored at - 5 °C. Prior to analysis of nutrients by ion chromatography samples were filtered over a 0.45 µm filter (mixed cellulose esters). Nitrate and phosphate concentrations were determined using an IC compact 761 (Metrohm, Herisau, Switzerland). The compounds were separated with a Metrosepp A SUPP5-column and determined with a conductivity detector. The detection limit of nitrate and phosphate was 0.01 mg/L.

5.3.6 Macrophytes

Three submersed macrophytes, the dicot *Myriophyllum spicatum* L. and the two monocots *Potamogeton lucens* L. and *Elodea canadensis* Michx., were selected as test species. *Myriophyllum spicatum* and *P. lucens* were both supplied by commercial nurseries, i.e. they were delivered by Alfred Forster AG (Golaten, Switzerland) and Lehnert Erb AG (Rombach, Switzerland), respectively, three weeks prior to application. *Elodea canadensis* was harvested from the mesocosms of the test site.

Two 7 cm cuttings of *M. spicatum* were planted together in one plastic pot of 5 cm diameter filled with rock wool (artificial substrate) to assure robust fixation. The same procedure was performed with *E. canadensis*. *Potamogeton lucens* was delivered as potted plants. The substrate for *P. lucens* comprised 28 % pumice, 24 % fractional expanded clay, 8 % sand, 12 % clay, 16 % bark humus and 12 % white peat. The plants had not developed any vegetative parts in early spring (April). In each mesocosm, twelve pots with *M. spicatum,* four pots with *E. canadensis* (four pots to measure length only) and ten pots with *P. lucens* were arranged on the top of plastic boxes (Semadeni, Ostermundigen, Switzerland) which were than introduced into the mesocosms. The macrophytes were posed in a depth of 70 cm below the surface and allowed to acclimate for three (*P. lucens, M. spicatum*) and two weeks (*E. canadensis*).

5.3.7 Biological endpoints

Photosynthetic efficiency

Effects on photosynthetic efficiency (PE) of the three submersed macrophytes were determined with a MINI PAM fluorometer (Heinz Walz GmbH, Effeltrich, Germany) by measuring *in vivo* chlorophyll fluorescence using the saturating pulse method (Schreiber 2004, Fai et al. 2007, Marwood et al. 2001). For that purpose, plants were lifted out of the pond. Subsequently, a leaf in the case of *P. lucens* or a shoot in the case of *M. spicatum* and *E. canadensis* was placed in the leaf clip holder of the PAM device. Ten measurements on different leaves or shoots were performed for each species always at the same time of the day on sampling days -1, 2, 5, 12, 19, 26, and 34 in the various treatments and control. PE determined as PSII quantum yield Y was calculated according to the following equation:

$$Y = (F_m' - F)/F_m'$$

where F is the present fluorescence yield measured between the pulses and F_m' is the maximum fluorescence yield induced by the saturation pulse leading to full inhibition of energy conversion at PSII reaction centers. For each macrophyte and each sampling day, PE values of the treated macrophytes were normalized to PE values obtained from the macrophytes in the control mesocosms.

Growth

To determine growth rates of *M. spicatum* and *E. canadensis*, lengths of the main and the side shoots were measured on day -1, 5, 12, 19, 26 and 34. Total plant length values, i.e. the sum of main and side shoots, were transformed to ln(length) values and plotted versus time to calculate relative growth rates (RGR) according to a least square linear regression.

5.3.8 Statistical analysis

If not otherwise mentioned, significance testing was generally performed using a one way ANOVA followed by Dunnett`s or Tukey`s test. The significance level was $p = 0.05$. To determine the variability of the measurements the coefficient of variation (CV) was calculated as:

$$\text{standard deviation/mean} \times 100\ \% = CV\ [\%]$$

5.4 Results

5.4.1 Physical/chemical parameters and nutrients

During the experiment, the temperature ranged between 14 and 18 °C and was similar in all mesocosms (Fig. 5.1). The functional endpoints pH, dissolved oxygen and conductivity indirectly indicated effects on photosynthesis efficiency of the mesocosm system including phytoplankton, periphyton and submerse macrophytes. Prior to application (day -1) when circulation was still running, values of pH (ranging between 7.8 and 8.3), dissolved oxygen (ranging between 104 and 125 %) and conductivity (ranging between 253 and 280 µS/cm) were comparable in all mesocosms (Fig. 5.1). After application, pH increased from 8.0 to 10.2 in the controls. Due to the herbicide effects, pH increase was slower and values were in general lower compared to the control. Results obtained from dissolved oxygen and conductivity measurements resulted in similar trends as described for the pH (Fig. 5.1). Nitrate concentrations in the water ranged between 0.01 and 0.25 mg/L whereby nitrate values were under the quantification limit of 0.01 mg/L in twelve and six ponds on day -23 and 16, respectively. Phosphate concentrations were in the range of 0.03 and 0.12 mg/L.

5.4.2 Chemical analysis

Mean water herbicide concentrations in the mesocosms determined in the single and the mixture treatments are presented in Fig. 5.2. Re-application of the substances was performed up to three times to maintain the target concentration. Atrazine was supplemented once on day 20 and isoproturon three times on day 12, 20 and 29 (Fig. 5.2). Diuron was re-applied twice on day 12 and 20. In the mixture, all herbicides were individually three times re-applied on day 12, 20 and 29. Averaged atrazine, isoproturon and diuron concentrations calculated time weighted average were in the range of the target concentration (Table 5.1). On two single exposure days (day 0 and day 12), atrazine (mixture treatment) and isoproturon (single treatment) were above and below the aimed target concentration ± 20 % in two and three ponds on day 0 and 12, respectively.

Chapter 5 - Phytotoxicity to submersed macrophytes

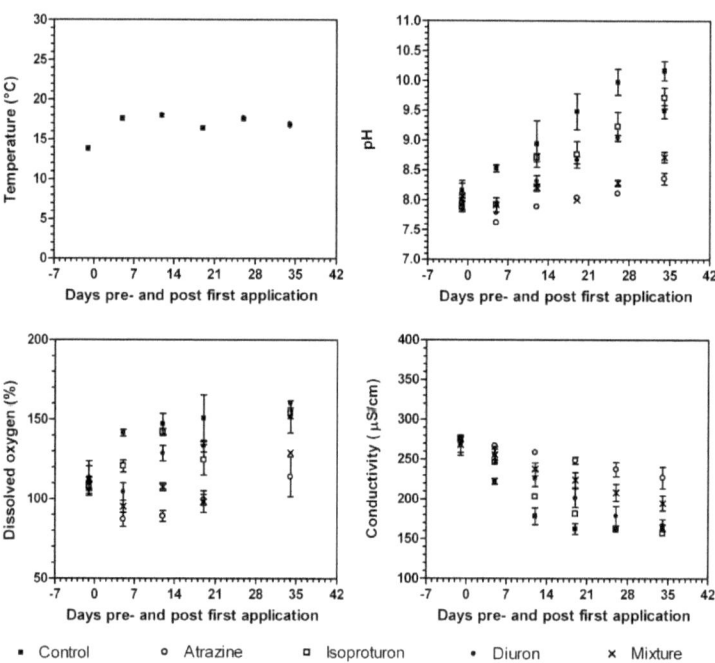

Fig. 5.1 Temperature, pH, dissolved oxygen, and conductivity measured for the four different treatments. Each data point indicates mean ± s.d. of three replicated ponds.

Table 5.1 Herbicide concentrations ($\mu g \cdot L^{-1}$) calculated as time weighted average for the single and mixture treatments. Mean ± s.d. are presented.

	Atrazine ($\mu g \cdot L^{-1}$)	Isoproturon ($\mu g \cdot L^{-1}$)	Diuron ($\mu g \cdot L^{-1}$)
Single treatment	76.1 ± 4.5	13.4 ± 2.4	4.9 ± 3.8
Mixture treatment	25.1 ± 3.1	4.6 ± 0.8	1.7 ± 0.2

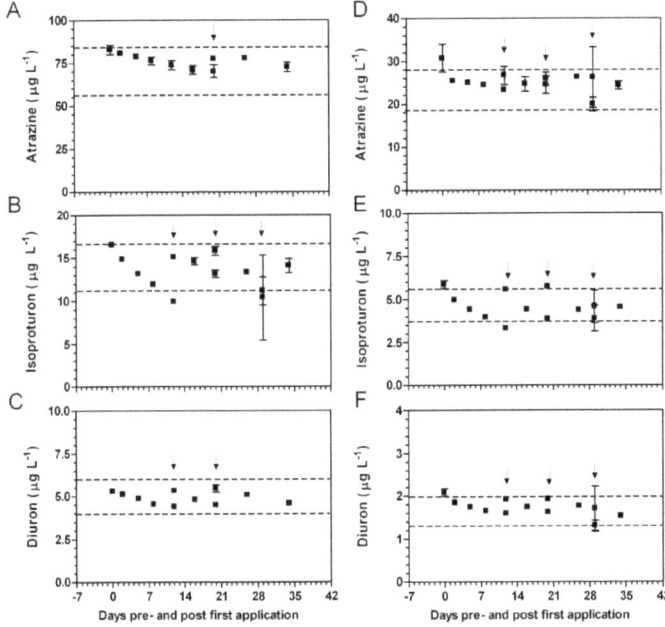

Fig. 5.2 Measured herbicide concentrations in µg/L for the different treatments, namely atrazine (A), isoproturon (B), diuron (C), atrazine in the mixture (D), isoproturon in the mixture (E) and diuron in the mixture (F). The dashed lines indicate the target concentration ± 20 %. Arrows indicate re-application of the herbicides. Each data point represents mean ± s.d. of three replicated ponds. The scales of the y-axes vary between the single diagrams A to F.

5.4.3 Biological endpoints

Photosynthetic efficiency

Prior to application, the PE of the macrophytes was not significantly different in all mesocosms (Fig. 5.3). After application, the PE was inhibited resulting in minimum values on day 2 to 5. On day 12, PE of the three submersed macrophytes corresponded to control values in all treatments despite further constant herbicide exposure.

Comparing species sensitivity, the dicotyle *Myriophyllum spicatum* was found to be most sensitive since all four treatments significantly reduced PE up to 30 to 60 % of the control (Fig. 5.4). The two monocots *E. canadensis* and *P. lucens* were found to be less sensitive. Atrazine significantly inhibited PE of these two macrophytes on day 5 only (Fig. 5.4). Diuron and the

Chapter 5 - Phytotoxicity to submersed macrophytes

mixture significantly inhibited PE of *E. canadensis* and *P. lucens* on days 2 and 5 and on day 5, respectively. PE of *E. canadensis* and *P. lucens* was not significantly affected by isoproturon.

Moreover, effects on PE were investigated with respect to mixture toxicity. On day 2 and day 5, PE of *M. spicatum* was comparably inhibited by the three single herbicides resulting in PE values of approximately 40 %. The herbicide mixture reduced of PE of *M. spicatum* similar to the single substances (one way ANOVA, p<0.05) (Fig. 5.5). Thus, the herbicides acted concentration additive in *M. spicatum* with respect to their effects on PE. In *E. canadensis* and *P. lucens* equitoxicity of the three single herbicide concentrations to PE could not be shown.

Concerning the test method, variability of *in vivo* chlorophyll fluorescence measurements was low and comparable between the three macrophytes. CVs corresponded to 19.4 ± 15.2 % (n = 35) in *E. canadensis*, 17.0 ± 14.4 % (n = 35) in *M. spicatum*, 19.4 ± 11.5 % (n =35) in *P. lucens*.

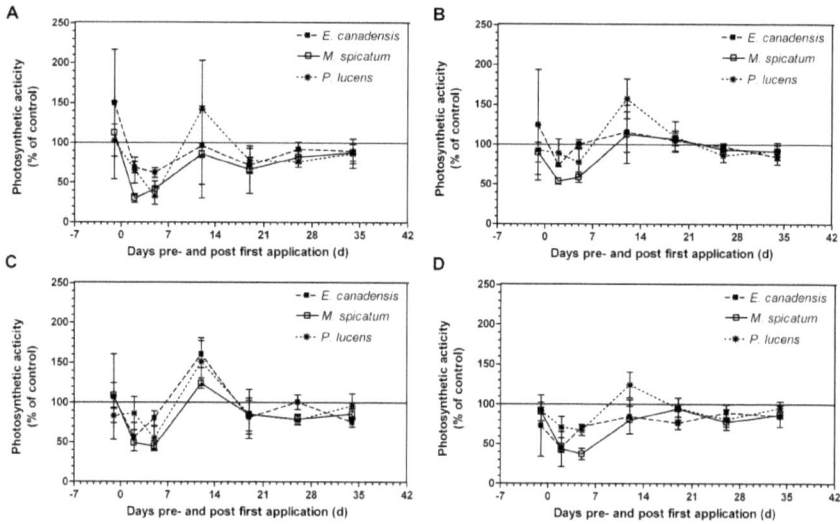

Fig. 5.3 PE (expressed as % of control) of the three submersed macrophytes *E. canadensis*, *M. spicatum* and *P. lucens* over time is shown for the four different treatments: atrazine (A), isoproturon (B), diuron (C) and the mixture (D). The horizontal line indicates the PE level corresponding to 100 % of control. Each data point represents mean ± s.d. of three replicated ponds.

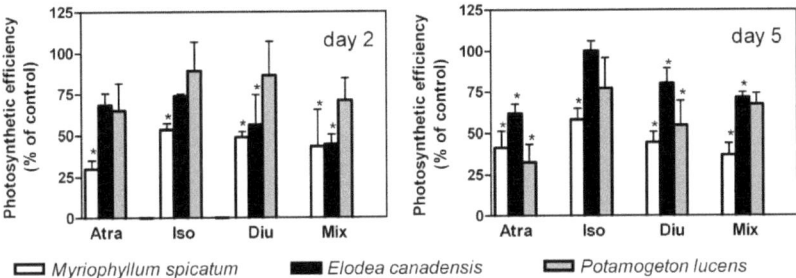

Fig. 5.4 PE of the three submersed macrophytes *E. canadensis, M. spicatum* and *P. lucens* determined on day 2 and 5 for different treatments. Bars represent mean ± s.d. of three replicated ponds. Asterisks above the bars indicate significant differences to PE values of the respective control between PE of the three macrophytes (one way ANOVA followed by Dunnett's test; $p<0.05$). Atra = Atrazine, Iso = Isoproturon, Diu = Diuron, Mix = Mixture.

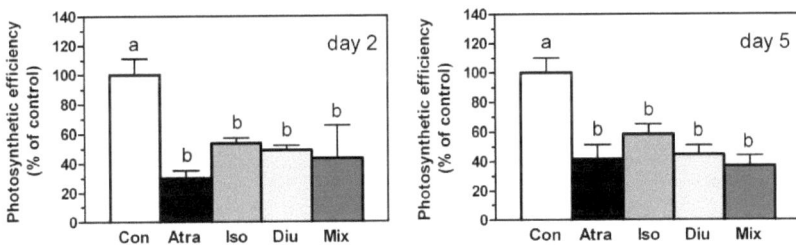

Fig. 5.5 PE of the submersed macrophyte *M. spicatum* determined on day 2 and 5 for different treatments. Bars represent mean ± s.d. of three replicated ponds. Bars not sharing a common letter (a, b) were found to be significantly different from each other (one way ANOVA followed by Tukey's multiple comparison test; $p<0.05$). Con = Control, Atra = Atrazine, Iso = Isoproturon, Diu = Diuron, Mix = Mixture.

Growth

Calculated RGR based on total length measurements of *M. spicatum* and *E. canadensis* are listed in Table 5.2. Neither growth of the dicotyle *M. spicatum* nor of the monocotyle *E. canadensis* was statistically significant reduced in the single as well as in the mixture treatment compared to the control. Variability of growth measurements was higher in *M. spicatum* compared to *E. canadensis* (Table 5.2).

Table 5.2 Relative growth rates (RGR) (based on total length) of *E. canadensis* and *M. spicatum* determined for the control, single herbicide and mixture treatment in the exposure period (day -1 – 34). RGR are expressed in % of control for all treatments. Values are means ± s.d. (n = 3). Growth of exposed macrophytes was not significantly reduced compared to those of the control treatment (one way ANOVA; p>0.05). Coefficients of variation (%) are given in brackets.

		E. canadensis	*M. spicatum*
RGR (% of control)	Control	100.0 ± 26.4	100.0 ± 50.8
		(26.4)	(50.8)
	Atrazine	113.3 ± 13.6	103.9 ± 24.0
		(12.0)	(23.1)
	Isoproturon	116.5 ± 5.8	119.8 ± 39.1
		(5.0)	(32.7)
	Diuron	111.6 ± 11.9	98.4 ± 25.0
		(10.6)	(25.4)
	Mixture	106.9 ± 15.7	101.2 ± 10.3
		(14.7)	(10.2)

5.5 Discussion

This study investigated the effects of environmental relevant concentrations of atrazine, isoproturon, diuron and of their mixture on photosynthesis of *E. canadensis, M. spicatum* and *P. lucens* and growth of *E. canadensis* and *M. spicatum* in outdoor mesocosms. Based on the results obtained from photosynthesis measurements, it was found that the dicotyle *M. spicatum* was more sensitive compared to the two monocotyles *E. canadensis* and *P. lucens*. Differences in the sensitivity might be explained by different metabolism pathways of the herbicides in monocotyledonous and dicotyledonous plants as demonstrated for isoproturon by Glässgen et al. (1999). However, adverse effects on photosynthesis could be observed in a short time window after application only. The rapid adaptation of the macrophytes to the chemical stressors resulted in no changes of plant growth at least of *E. canadensis* and *M. spicatum*.

Despite the lack of comparative laboratory data regarding the toxicity of isoproturon and diuron to *E. canadensis* and isoproturon to *M. spicatum* these two macrophytes were mostly found to be more sensitive to these PSII inhibitors in laboratory compared to field studies. Growth of *E. canadensis* determined as biomass and shoot length, for instance, was shown to be not significantly affected by atrazine concentrations up to 75 µg/L in flow-through wetland mesocosms which is in agreement with our observations (Detenbeck et al. 1996). In contrast, the EC_{50} based on laboratory measurements of wet weight after 14 d exposure to atrazine was determined to be 21 µg/L for *E. canadensis*, a concentration 3 fold lower than the concentration in our mesocosms (Fairchild et al. 1998). Further laboratory studies demonstrated a 50 % reduction in shoot growth of *E. canadensis* after exposure to 80 µg atrazine/L for 28 days and 163 µg atrazine/L for 42 days, respectively (Forney and Davis 1981). Sensitivity of *M. spicatum* to atrazine in the laboratory and field studies appeared to be comparable since in the present mesocosm study no significant effects of 70 µg atrazine/L on shoot growth were found and the EC_{50} based on shoot growth after 28 days was 1104 µg/L in laboratory experiments (Forney and Davis 1981). However, *M. spicatum* was more sensitive to diuron under laboratory than under field conditions. In our study 5 µg diuron/L exerted no adverse effects on *M. spicatum* shoot growth whereas the same concentration induced a 50 % reduction in *M. spicatum* biomass measured as dry weight after 14 days of exposure (Lambert et al. 2006). That herbicides have often stronger effects on fast compared to slowly growing plants might be an explanation for the observed higher sensitivities of *M. spicatum* and *E. canadensis* towards PSII herbicides in laboratory compared to field studies. Nutrient supply and light conditions are mostly not limited in laboratory experiments resulting in optimal growth of macrophytes, whereas macrophytes and algae often compete for light and nutrients in the field.

Regarding the time course of the functional endpoints pH, dissolved oxygen and conductivity in the herbicide treated mesocosms, photosynthetic performance of the entire model ecosystem still appeared to be impaired at the end of the five week exposure period. This observation might be best explained by the contribution of phytoplankton and periphyton to the photosynthetic performance of the mesocosms. The delay of pH recovery compared to the recovery of macrophyte photosynthesis thus might be partly explained by an observed 40 % inhibition of phytoplankton photosynthesis in the single and mixture herbicide treatments during the five week exposure period (Knauert et al. 2008).

Moreover, our results revealed that photosynthetic efficiency, an endpoint directly linked to the mode of action of the herbicides was useful to examine the applicability of concentration addition in macrophytes under semi-field conditions. Atrazine, isoproturon, and diuron were shown to act concentration additive in *M. spicatum* concerning their effects on photosynthesis. For *E. canadensis* and *P. lucens*, however, it was difficult to draw any conclusions on mixture toxicity since the single herbicide concentrations did not induce comparable effects. However, results obtained from *M. spicatum* are in consistence with findings of previous laboratory studies demonstrating that the concept of concentration addition is a useful tool to predict the joint toxicity of similar acting PSII herbicides (triazines and phenylureas) in the green algae *Scenedesmus vacuolatus* as well as in the aquatic macrophyte *Lemna minor* (Faust et al. 2001, Drost et al. 2003, Backhaus et al. 2004). In the light of these results, we recommend taking concentration addition into consideration when monitoring and regulating water quality with respect to herbicide contamination. A possible approach has been recently proposed by Chèvre et al (2006).

In vivo chlorophyll fluorescence as a mean to determine effects on photosynthetic efficiency of macrophytes turned out to be a sensitive and thus valuable biomarker to monitor PSII herbicide contamination in the aquatic environment. However, plant growth remains the more relevant ecological endpoint when assessing the ecotoxicological risk of herbicides for the aquatic environment since mainly reduction in plant growth will have a serious impact on the ecological functioning of aquatic ecosystems. Taking this into account, we conclude from our experiment that a constant exposure to environmental relevant concentrations of atrazine, isoproturon, and diuron and to an environmental relevant mixture of these three compounds did not pose a risk to *M. spicatum* and *E. canadensis*.

5.6 References

Backhaus, T., Faust, M., Scholze, M., Gramatica, P., Vighi, M., Grimme, L. H., 2004. Joint algal toxicity of phenylurea herbicides is equally predictable by concentration addition and independent action. *Environmental Toxicology and Chemistry* 23, 258-264.

Blanchoud, H., Farrugia, F., Mouchel, J.M., 2004. Pesticide uses and transfers in urbanised catchments. *Chemosphere* 55, 905-913.

Carpenter, S. R., Lodge, D. M., 1986. Effects of submersed macrophytes on ecosystem processes. *Aquatic Botany* 26, 341-370.

Chèvre, N., Loeppe, C., Singer, H., Stamm, C., Fenner, K., Escher, B. I., 2006. Including mixtures in the determination of water quality criteria for herbicides in surface water. *Environmental Science & Technology* 40, 426-435.

Detenbeck, N. E., Hermanutz, R., Allen, K., Swift, M.C., 1996. Fate and effects of the herbicide atrazine in flow-through wetland mesocosms. *Environmental Toxicology and Chemistry* 15, 937-946.

Drost, W., Backhaus, T., Vassilakaki, M., Grimme, L. H., 2003. Mixture toxicity of s-triazines to *Lemna minor* under conditions of simultaneous and sequential exposure. *Fresenius Environmental Bulletin* 12, 601-607.

Fai, P.B., Grant, A., Reid, B., 2007. Chlorophyll a fluorescence as a biomarker for rapid toxicity assessment. *Environmental Toxicology and Chemistry* 26, 1520-1531.

Fairchild, J. F., Lapoint, T. W., Schwartz, T. R., 1994. Effects of an herbicide and insecticide mixture in aquatic mesocosms. *Archives of Environmental Contamination and Toxicology* 27, 527-533.

Fairchild, J. F., Ruessler, D. S., Carlson, A. R., 1998. Comparative sensitivity of five species of macrophytes and six species of algae to atrazine, metribuzin, alachlor, and metolachlor. *Environmental Toxicology and Chemistry* 17, 1830-1834.

Faust, M., Altenburger, R., Backhaus, T., Blanck, H., Boedeker, W., Gramatica, P., Hamer, V., Scholze, M., Vighi, M., Grimme, L. H., 2001. Predicting the joint algal toxicity of multi-component s-triazine mixtures at low-effect concentrations of individual toxicants. *Aquatic Toxicology* 56, 13-32.

Feurtet-Mazel, A., Grollier, T., Grouselle, M., Ribeyre, F., Boudou, A., 1996. Experimental study of bioaccumulation and effects of the herbicide isoproturon on freshwater rooted macrophytes - (*Elodea densa* and *Ludwigia natans*). *Chemosphere* 32, 1499-1512.

Field, J.A., Reed, R.L., Sawyer, T.E., Griffith, S.M., Wigington, P.J., 2003. Diuron occurrence and distribution in soil and surface and ground water associated with grass seed production. *Journal of Environmental Quality* 32, 171-179.

Forney, D. R, Davis, D. E., 1981. Effects of low concentrations of herbicides on submersed aquatic plants. *Weed Science* 29, 677-685.

Freitas, L. G., Götz, C.W., Ruff, M., Singer, H.P., Müller, S.R., 2004. Quantification of the new triketone herbicides, sulcotrione and mesotrione, and other important herbicides and metabolites, at the ng/l level in surface waters using liquid chromatography-tandem mass spectrometry. *Journal of Chromatography A* 1028, 277-286.

Garmouma, M., Chevreuil, M., 1998. Triazine dispersion and distribution in the unsaturated zone of drained soils in the Brie (France). *Water Air and Soil Pollution* 108, 129-148.

Gfrerer, M., Martens, D., Gawlik, B. M., Wenzl, T., Zhang, A., Quan, X., Sun, C., Chen, J., Platzer, B., Lankmayr, E., Kettrup, A., 2002. Triazines in the aquatic systems of the Eastern Chinese Rivers Liao-He and Yangtse. *Chemosphere* 47, 455-466.

Gilliom, R. J., Barbash, J. E., Kolpin, D. W., Larson, A. J., 1999. Testing water quality for pesticide pollution. *Environmental Science & Technology* 33, 164A-169A.

Gläßgen, W.E., Komoßa, D., Bohnenkämper, O., Haas, M., Hertkorn, N., May, R.G., Szymczak, W., Sandermann, H., 1999. Metabolism of the herbicide isoproturon in wheat and soybean cell suspension cultures. *Pesticide Biochemistry and Physiology* 63, 97-113.

Graymore, M., Stagnitti, F., Allinson, G., 2001. Impacts of atrazine in aquatic ecosystems. *Environment International* 26, 483-495.

Huang, X. J., Pedersen, T., Fischer, M., White, R., Young, T. M., 2004. Herbicide runoff along highways. 1. Field observations. *Environmental Science & Technology* 38, 3263-3271.

Huber, W., 1993. Ecotoxicological Relevance of Atrazine in Aquatic Systems. *Environmental Toxicology and Chemistry* 12, 1865-1881.

Irace-Guigand, S., Aaron, J.J., Scribe, P., Barcelo, D., 2004. A comparison of the environmental impact of pesticide multiresidues and their occurrence in river waters surveyed by liquid chromatography coupled in tandem with UV diode array detection and mass spectrometry. *Chemosphere* 55, 973-981.

Key, P., Chung, K., Siewicki, T., Fulton, M., 2007. Toxicity of three pesticides individually and in mixture to larval grass shrimp (*Palaemonetes pugio*). *Ecotoxicology and Environmental Safety* 68, 272-277.

Kirby, M.F., Sheahan, D.A., 1994. Effects of Atrazine, Isoproturon, and Mecoprop on the Macrophyte *Lemna minor* and the alga *Scenedesmus subspicatus*. *Bulletin of Environmental Contamination and Toxicology* 53, 120-126.

Knauer, K., 1993. *Natural and artificial shores at Lake Constance - a comparison of littoral biocoenosis in front of natural reed shores and man made walls at the lakeside*, in: Lakeshore Deterioration and Restoration Works in Central Europe, Limnologie Aktuell Band 5, eds. Ostendorp, W., Krumscheid-Plankert, P, Gustav Fischer Verlag, Stuttgart, Germany, pp. 189-195.

Knauer, K., Maise, S., Thoma, G., Hommen, U., Gonzalez-Valero, J., 2005. Long-term variability of zooplankton populations in aquatic mesocosms. *Environmental Toxicology and Chemistry* 24, 1182-1189.

Knauer, K., Vervliet-Scheebaum, M., Dark, R. J., Maund, S. J., 2006. Methods for assessing the toxicity of herbicides to submersed aquatic plants. *Pesticide Management Science* 62, 715-722.

Knauert, S., Escher, B., Singer, H., Hollender, J., Knauer, K., 2008. Mixture toxicity of three photosystem II inhibitors (atrazine, isoproturon, and diuron) towards photosynthesis of freshwater phytoplankton studied in outdoor mesocosms. *Environmental Science& Technology*

Körner, S., 2002. Loss of submerged macrophytes in shallow lakes in North-Eastern Germany. *International Review of Hydrobiology* 87, 375-384.

Kreuger, J., 1998. Pesticides in stream water within an agricultural catchment in southern Sweden, 1990-1996. *Science of the Total Environment* 216, 227-251.

Lambert, S. J., Thomas, K. V., Davy, A. J., 2006. Assessment of the risk posed by the antifouling booster biocides Irgarol 1051 and diuron to freshwater macrophytes. *Chemosphere* 63, 734-743.

Lewis, M. A., 1995. Use of freshwater plants for phytotoxicity testing: A review. *Environmental Pollution* 87, 319-336.

Lytle, J. S., Lytle, T. F., 2002. Uptake and loss of chlorpyrifos and atrazine by *Juncus effusus* L. in a mesocosm study with a mixture of pesticides. *Environmental Toxicology and Chemistry* 21, 1817-1825.

Lytle, T. F., Lytle, J. S., 2005. Growth inhibition as indicator of stress because of atrazine following multiple toxicant exposure of the freshwater macrophyte, *Juncus effusus* L. *Environmental Toxicology and Chemistry* 24, 1198-1203.

Marwood, C.A., Solomon, K.R., Greenberg, B.M., 2001. Chlorophyll fluorescence as a bioindicator of effects on growth in aquatic macrophytes from mixtures of polycyclic aromatic hydrocarbons. *Environmental Toxicology and Chemistry* 20, 890-898.

Sand-Jensen, K., Riis, T., Vestergaard, O., Larsen, S. E., 2000. Macrophyte decline in Danish lakes and streams over the past 100 years. *Journal of Ecology* 88, 1030-1040.

Schreiber, U., 2004. *Pulse-Amplitude-Modulation (PAM) fluorometry and saturation pulse method: An overview*, in: Chlorophyll a Fluorescence: A Signature of Photosynthesis, eds. Papageorgiou, G.C., Govindjee, R, Kluwer Academic Publishers, Dordrecht, The Netherlands, pp. 279-319.

Snel, J. F. H., Vos, J. H., Gylstra, R., Brock, T. C. M., 1998. Inhibition of photosystem II (PSII) electron transport as a convenient endpoint to assess stress of herbicide linuron on freshwater plants. *Aquatic Ecology* 32, 113-123.

Solomon, K. R., Baker, D. B., Richards, R. P., Dixon, D. R., Klaine, S. J., LaPoint, T. W., Kendall, R. J., Weisskopf, C. P., Giddings, J. M., Giesy, J. P., Hall, L. W., Williams, W. M., 1996. Ecological risk assessment of atrazine in North American surface waters. *Environmental Toxicology and Chemistry* 15, 31-74.

Teisseire, H., Couderchet, M., Vernet, G., 1999. Phytotoxicity of diuron alone and in combination with copper or folpet on duckweed (*Lemna minor*). *Environmental Pollution* 106, 39-45.

Trebst, A., 1987. The 3-dimensional structure of the herbicide binding niche on the reaction center polypeptides of photosystem-II. *Zeitschrift für Naturforschung C - Journal of Biosciences* 42, 742-750.

Wetzel, R.G., 2001. *Limnology - Lake and River ecosystems*. 3rd edition, Academic Press, San Diego, CA, USA.

Chapter 6

Concluding remarks and future directions

6.1. Challenges in regulating pesticide mixtures

Most ecotoxicological studies focus on effects of single stressors on ecosystem components. However, organisms in the environment are often exposed to many stressors simultaneously, including those of a physical, biological, and chemical nature. Regulatory methods for the management of chemical compounds are mostly based on risk evaluations of single-substances. Recently, effects of mixtures are evaluated with toxicological models to predict the joint effect on single species. The extrapolation from single-species mixture toxicity to *in situ* risk for natural communities exposed to mixtures of pollutants adds complexity, but is a challenge which has to be taken in the near future.

In Chapter 3 of this thesis, the HC_{30} of atrazine, isoproturon, and diuron were found to be equitoxic to phytoplankton communities investigating photosynthesis in laboratory and field experiments. Photosynthesis was chosen as toxicological endpoint since it is directly linked to the mode of action of the test substances. It was further demonstrated that the effects of the PSII herbicide mixture can be explained by concentration addition of the individual components.

In Chapter 4 of this thesis, the HC_{30} of atrazine, isoproturon, and diuron and their mixture were found to stimulate comparable effects on community level expressed as total abundance and diversity of phytoplankton despite the different sensitivities of single algal species towards the various treatments. However, in the long term, the community structure of phytoplankton developed differently from each other due to differences in the sensitivity of a few algal species and due to the different dissipation of the herbicides. Recovery of the community was assumed to be partly linked to the dissipation of the substances and to enhanced tolerance of the algae towards the herbicides. Pollution induced community tolerance was investigated during the present project by a trainee and will be published separately.

The findings of the chapters 3 & 4 gave evidence that the concept of concentration addition also applies for the prediction of mixture effects under environmental conditions. Further studies should be performed with other groups of pesticides to confirm these results for other toxic modes of action involving other organisms and toxicological endpoints. If the concept of concentration addition holds true also for other substances, it can be recommended assessing the risk of pesticide mixtures based on the sum of the measured environmental concentrations and of related toxicological effects. Thereby it might be advisable to summarize pesticides according to their mode of action. Regarding herbicides, an overall risk based on mixture toxicity might be determined for the group of PSII inhibitors, inhibitors of branched chain amino acid biosynthesis, auxin simulators, and others. Regarding insecticides, carbamates, and organophosphates could be summarized as one group since these chemicals act as inhibitors of the enzyme acetylcholine esterase. Chlorinated organic hydrocarbons (e.g, dieldrin, lindane, DDT) and pyrethroids that specifically interact with neuronal sodium channels resulting in nervous system hyperexcitability in insects could constitute another group.

Assessing the risk of mixtures composed of chemicals from different classes with completely different modes of toxic action, e.g., insecticides with herbicides, remains a future challenge. Research on the impacts of such mixtures on organisms has yielded mixed results. For example, no toxic interaction was noted when *Ceriodaphnia tentans* were exposed to a binary mixture of the triazine atrazine and the carbamate insecticide carbofuran (Douglas et al. 1993). The joint toxicity of the organophosphate diazinon and ammonia was examined in *Ceriodaphnia dubia* with dosed water and effluents containing both stressors (Bailey et al. 2001). The results indicated a less than additive response for the binary mixture in both laboratory-dosed and effluent samples. In a separate study using *Ceriodaphnia dubia*, Banks et al. (2003) found less than additive responses for binary mixtures of diazinon and copper. Munkegaard et al. (2008) tested whether a potential inhibition of P450 by organophosphorous insecticides in aquatic plants and algae could increase their sensitivity to herbicide exposure. Their study showed no synergistic interactions between the insecticides malathion, endosulfan, and chlorpyrifos and the herbicides metsulfuron-methyl, terbutylazine, and bentazone in *Lemna minor* and *Pseudokirchneriella subcapitata*. Other studies reported on toxicity greater than additive responses. For instance, Key et al. (2007) demonstrated that a binary mixture of two insecticides fipronil and imidacloprid was not more toxic to larvae of grass shrimp (*Palaemonetes pugio*) than the insecticides alone. However, the addition of nontoxic concentrations of atrazine increased the toxicity of this binary mixture. Furthermore, triazine herbicides were demonstrated to enhance the effects of some organophosphates (Pape-Lindstrom and Lydy 1997, Lydy and Linck 2004). The authors suggested that atrazine increased the biotransformation of organophosphates by converting them into more toxic metabolites. Atrazine might be accomplishing the metabolic activation by inducing the cytochrome P450 enzymes responsible for the conversion. It is clear from these studies that toxicokinetic processes such as biotransformation have to be considered when determining the toxicity of some pesticide mixtures.

Aquatic ecosystems exhibit a myriad of physical and biological variables that would need to be included along with the chemical stressors to more accurately model real environmental exposure scenarios. For example, using a bacterial bioluminescence inhibition assay, Benson and Long (1991) found that humic acids significantly reduced the toxicity of acetylcholine esterase inhibitors, e.g., chlorpyrifos and carbofuran, while enhancing the toxicity of others such as methyl parathion and carbaryl. Temperature has been demonstrated to have an inverse effect on pyrethroid toxicity to larvae of *Culex quinquefasciatus say*, which increases at lower temperatures (Mahboob et al. 1999). Moreover, Talent (2005) showed that the sensitivity of green anole lizards to natural pyrethrin pesticides increased with decreasing temperature. The primary reason that toxicity of pyrethrins varies with the temperature of poikilothermic animals is because at lower temperatures, neurons are more susceptible to excitation as a result of increased sodium current flow through sodium channels (Narahashi 2001). Conversely, Lydy et al. (1990) reported increases as high as 100-fold in the toxicity of the organophosphate parathion at higher temperatures. Herbrandson et al. (2003a,b) examined the effects of suspended solids on carbofuran toxicity to *Daphnia magna* and found no measurable toxicity associated with exposure to suspended solids at a wide range of concentrations. However, when exposed to a constant concentration of carbofuran, the number of affected organisms increased with increasing concentration of suspended solids. The authors speculated that the ingestion of the solids was causing the *Daphnia magna* to sink, which forced them to expend significantly more energy to maintain proper buoyancy. In turn, this increased energy expenditure made the *Daphnia magna* more susceptible to carbofuran toxicity. Food availability can also affect the toxicity of a pesticide. Barry et al. (1995) demonstrated that esfenvalerate toxicity to *Daphnia carinata* increased significantly with decreasing food concentration. The converse is also possible in that increased food density can result in decreased toxicity. Herbrandson et al. (2003a,b) showed that increased food availability significantly reduced carbofuran toxicity to *Daphnia magna*. However, the mechanism behind these observations is difficult to interpret, because the observed effects could be the result of changes in either organism fitness or toxicant concentration caused by sorption to the food source. Taken together, attempts to perform toxicity studies that model realistic environmental exposure scenarios should account for variables such as temperature, pH, food availability, etc.

6.2 Challenges in protecting sustainable freshwater ecosystems

When assessing the aquatic risks of the agricultural use of pesticides, it is important to have a scientifically sound idea of what constitutes an important ecological effect of these chemicals in surface waters and what constitutes a sustainable freshwater ecosystem. To sustain thereby means to "hold" or to "keep alive". Sustainability of freshwater ecosystems concerns not only their ecological properties but also their economic and social functions. Within the context of the sustainability of freshwater communities, three general categories of effects of pesticides in the environment can be distinguished. They are related to ecosystem structure and functioning but also to aesthetic and economic values of humans (Calow 1998, Brock and Ratte 2002).

The structure of an ecosystem is a combination of which and how many organisms are present. Changes in structure generally are expressed in terms of overall species richness and densities as well as population densities of key and indicator species. Ecological key species are species that play a major role in ecosystem performance, productivity, stability, and resilience. Their impact on the other components of the community often is disproportionally large relative to their abundance (Power et al. 1996). It may concern species that are critical determinants in trophic cascades (e.g., top predators) or species with properties of ecological engineers, having great impact on the physical properties of the habitat, such as macrophytes. In this context, it is important to realize that the concern rarely is for individual organisms but, rather, for the viability of the total population in the habitat of concern. Ecological indicator species concern species with high information content for monitoring purposes, species protected by law, or regionally rare or endangered species.

In the present study, adverse effects on photosynthesis of phytoplankton resulted in structural changes of the aquatic ecosystem since the total abundance of phytoplankton decreased (see Chapter 4). Herbicide impact on periphyton has not been investigated and thus cannot be judged with respect to structural changes. Overall, the reduction in the algal population might impact higher levels in the food web of the aquatic ecosystem. Due to less food supply, the zooplankton population and grazing organisms were expected to decrease in their abundance. On the other hand, reduced photosynthesis of macrophytes did not lead to structural effects since growth of the macrophytes was not impaired (Chapter 5). To get a better understanding, which processes contributed to the rapid adaptation of the macrophytes to the herbicide impact, further investigations are needed. A rapid adaptation to herbicide stress has been reported to involve e.g., the activation of the biotransformation and detoxification systems (cytochrome P450, gluthathion-S-transferases and peroxidases). An ongoing collaboration with the University of Konstanz is addressing the role of polyphenols as possible scavengers of herbicide-induced reactive oxygen species. This research project aims to answer the questions whether an induction of polyphenols can be observed in the plant samples and whether an increase in the polyphenol content can be linked to the herbicide exposure.

The function of an ecosystem relates to what the organisms do in the ecosystem. Impact on ecosystem functioning concerns negative effects on biogeochemical cycles and energy flow.

According to the ecological theory, the protection of community structure will ensure the maintenance of ecosystem function (Levine 1989), whereas the loss of certain species in the community does not necessarily affect ecosystem processes because of the redundancy in roles and functions of the surviving species (Lawton 1994). Key is the preservation of the ecosystem's capacity to keep on functioning, and to keep the ability for self-regulation, under a range of environmental conditions that fall within the normal operating range of the system (Swift et al. 2004).

Community recovery occurs when an ecosystem can adsorb and endure a certain amount of pollution because of ecological recovery processes. The stressor should be limited to an intensity or concentration that causes reversible impacts only on the most sensitive populations. From a scientific point of view, periodically occurring declines in population densities can be considered as a normal phenomenon in ecosystems. In the course of their evolution, organisms have developed a large variety of strategies to survive and cope with temporally variable and unfavorable conditions, such as desiccation, flooding, temperature shocks, shading, oxygen depletion, and anthropogenic stressors (Ellis 1989). In some cases the stress caused by a pesticide may more or less resemble that of a natural stress factor. The use of the "normal operating range" of population densities and functional endpoints in specific ecosystems has been suggested as a baseline against which to assess pesticide induced changes (Domsch et al. 1983). In other words, effects of pesticides for which the bioavailable fraction is restricted in space and time may, in certain habitats, be regarded as ecologically unimportant when they are of a smaller scale than changes caused by other natural or anthropogenic stresses. In the present study, community composition of the phytoplankton exposed to the fast dissipating phenyl ureas diuron and isoproturon recovered within a time of two weeks. Thus, the observed effects of these two treatments can be judged as ecologically unimportant. For slow dissipating compounds, such as atrazine, the effects continued and a recovery was not demonstrated during a five-month period implying a sustained damage to the aquatic community.

The EU Water Framework Directive aims to achieve a "good status" for European surface waters, which will be achieved, in part, by protecting the aquatic communities from chemical stress. Biomarkers can be used as monitoring tools for an early detection of a water contamination. The H_2DCFDA fluorescence based technique to directly determine oxidative stress in algae was shown to be a useful method to detect effects of environmental relevant copper concentrations on green algal cells in laboratory experiments (Chapter 2). However, this technique did not enable the detection of oxidative stress induced by environmental relevant concentrations of the three herbicides atrazine, isoproturon, and diuron (data not shown). The endpoint photosynthetic activity revealed to be a sensitive and thus valuable biomarker to monitor an exposure to environmental relevant concentrations of the PSII inhibitors atrazine, isoproturon, and diuron as well as of copper in laboratory and field experiments. However, a biomarker response such as the induction of oxidative stress or the inhibition of photosynthetic activity cannot be used to judge the ecological relevance of an environmental contamination. Only endpoints that are relevant at the organisms or population level such as species richness, reproduction, growth, and abundance of key species are relevant to assess the potential risk of pesticides and are thus used to derive environmental quality criteria.

The decision to allow the agricultural use of pesticides implies the acceptance of effects on target organisms and, inevitably, the acceptance of certain effects on nontarget organisms in the agroecosystem as well. Only society can decide what levels of disturbance to freshwater ecosystems will be acceptable or not. Human demands for economic benefits from these ecosystems strongly affect this decision. There also may be disagreements between different sectors of society as to what constitutes a desirable benefit (Frewer 1999). An important role of scientists in this acceptability debate is to act as informers proposing decision schemes and presenting options for deriving acceptable pesticide concentrations in surface waters.

6.3 References

Bailey, H.C., Elphick, J.R., Krassoi, R., Lovell, A., 2001. Joint acute toxicity of diazinon and ammonia to *Ceriodaphnia dubia*. *Environmental Toxicology and Chemistry* 20, 2877-2882.

Banks, K., Wood, S., Matthews, C., Theusen, K., 2003. Joint acute toxicity of diazinon and copper to *Ceriodaphnia dubia*. *Environmental Toxicology and Chemistry* 22, 1562-1567.

Barry, M.J., Logan, D.C., Ahokas, J.T., Holdway, D.A., 1995. Effect of algal food concentration on toxicity of two agricultural pesticides to *Daphnia carinata*. *Ecotoxicology and Environmental Safety* 32, 273-279.

Benson, W.H., Long, S.F., 1991. Evaluation of humic-pesticide interactions on the acute toxicity of selected organophosphate and carbamate insecticides. *Ecotoxicology and Environmental Safety* 21, 301-307.

Brock, T.C.M., Ratte, H.T., 2002. *Ecological risk assessment for pesticides*, in: Community-level aquatic system studies – Interpretation criteria, eds. Giddings, J.M., Brock, T.C.M., Heger, W., Heimbach, F., Maund, S.J., Norman, S.M., Ratte, H.J., Schäfers, C., Streloke, M., SETAC press, Pensacola, FL, USA, pp. 33-41.

Calow, P., 1998. Ecological risk assessment: Risk for what? How do we decide? *Ecotoxicology and Environmental Safety* 40, 15-18.

Domsch, K.H., Jagnow, G., Anderson, T.H., 1983. An ecological concept for the assessment of side effects of agrochemicals on soil microorganisms. *Residue Reviews* 86, 65-105.

Douglas, W.S., McIntosh, A., Clausen, J.C., 1993. Toxicity of sediments containing atrazine and carbofuran to larvae of the midge *C. tentans. Environmental Toxicology and Chemistry* 12, 847-853.

Ellis, D., 1989. *Environments at risk: Case histories of impact assessment.* Springer Verlag, New York, NY, USA.

Frewer, L.J., 1999. Risk perception, social trust, and public participation into strategic decision-making – Implications for emerging technologies. *Ambio* 28, 569-574.

Herbrandson, C., Bradbury, S.P., Swackhamer, D.L., 2003a. Influence of suspended solids on acute toxicity of carbofuran to *Daphnia magna*. I. Interactive effects. *Aquatic Toxicology* 63, 333-342.

Herbrandson, C., Bradbury, S.P., Swackhamer, D.L., 2003b. Influence of suspended solids on acute toxicity of carbofuran to *Daphnia magna*. II. An evaluation of potential interactive mechanisms. *Aquatic Toxicology* 63, 343-355.

Key, P., Chung, K., Siewicki, T., Fulton, M., 2007. Toxicity of three pesticides individually and in mixture to larval grass shrimp (*Palaemonetes pugio*). *Ecotoxicology and Environmental Safety* 68, 272-277.

Lawton, J.H., 1994. What do species do in eosystems? *Oikos* 71, 367-374.

Levine, S.N., 1989. *Theoretical and methodological reasons for variability in the responses of aquatic ecosystem processes to chemical stress*, in: Ecotoxicology: Problems and approaches, eds. Levin, S.A., Harwell, M.A., Kelly, J.R., Kimball, K. D., Springer Verlag, New York, NY, USA, pp 145-179.

Lydy, M.J., Lohner, T.W., Fisher, S.W., 1990. Influence of pH, temperature and sediment type on the toxicity, accumulation, and degradation of parathion in aquatic systems. *Aquatic toxicology* 17, 27-44.

Lydy, M.J., Linck, S.L., 2004. Assessing the impact of triazine herbicides on organophosphate insecticide toxicity to the earthworm *Eisenia fetida. Archives of Environmental Contamination and Toxicology* 45, 343-349.

Mahboob, S.M., Howlader, A.J., Shahjahan, R.M., 1999. Effect of temperature on the toxicity of three insecticides against the fourth instar larvae of *Culex quinquefasciatus say* (Diptera: Culicidae). *Bangladesh Journal of Zoology* 27, 185-189.

Munkegaard, M., Abbaspoor, M., Cedergreen, N., 2008. Organophosphorous insecticides as herbicide synergists on the green algae *Pseudokirchneriella subcapitata* and the aquatic plant *Lemna minor*. *Ecotoxicology* 17, 29-35.

Narahashi, T., 2001. Recent progress in the mechanism of action of insecticides: Pyrethroids, fipronil, and indoxacarb. *Journal of Pesticide Science* 26, 277-285.

Pape-Lindstrom, P.A., Lydy, M.J., 1997. Synergistic toxicity of atrazine and organophosphate insecticides contravenes the response addition mixture model. *Environmental Toxicology and Chemistry* 16, 2415-2420.

Power, M.E., Tilman, D., Estes, J.A., Menge, B.A., Bond, W.J., Mills, L.S., Daily, G., Castilla, J.C., Lubchenco, J., Paine, R.T., 1996. Challenges in the quest for keystones. *Bioscience* 46, 609-620.

Swift, M.J., Izak, A.M.N., Van Noordwijk, M., 2004. Biodiversity and ecosystem services in agricultural landscapes – Are we asking the right questions? *Agriculture Ecosystem Environment* 104, 113-134.

Talent, L.G., 2005. Effect of temperature on toxicity of a natural pyrethrin pesticide to green anole lizards (*Anolis carolinensis*). *Environmental Toxicology and Chemistry* 24, 3113-3116.

Südwestdeutscher Verlag für Hochschulschriften

Wissenschaftlicher Buchverlag bietet
kostenfreie
Publikation
von
Dissertationen und Habilitationen

Sie verfügen über eine wissenschaftliche Abschlußarbeit zu aktuellen oder zeitlosen Fragestellungen, die hohen inhaltlichen und formalen Anspruchen genügt, und haben **Interesse an einer honorarvergüteten Publikation?**

Dann senden Sie bitte erste Informationen über Ihre Arbeit per Email an: info@svh-verlag.de.

Unser Außenlektorat meldet sich umgehend bei Ihnen.

Südwestdeutscher Verlag für Hochschulschriften
Aktiengesellschaft & Co. KG
Dudweiler Landstr. 99
D – 66123 Saarbrücken
www.svh-verlag.de

Printed by Books on Demand GmbH, Norderstedt / Germany